Food Processing

Food Processing

Advances in Thermal Technologies

Edited by

Kshirod Kumar Dash

Sourav Chakraborty

CRC Press
Taylor & Francis Group
Boca Raton London New York

CRC Press is an imprint of the
Taylor & Francis Group, an **informa** business

First edition published 2021 by
CRC Press
6000 Broken Sound Parkway NW, Suite 300, Boca Raton, FL 33487-2742

and by CRC Press
2 Park Square, Milton Park, Abingdon, Oxon, OX14 4RN

First edition published by CRC Press 2021

CRC Press is an imprint of Taylor & Francis Group, LLC

ISBN: 978-0-367-33720-9 (hbk)
ISBN: 978-0-367-69547-7 (pbk)
ISBN: 978-0-429-32148-1 (ebk)

Typeset in Times
by SPi Global, India

Contents

Preface

Advances in thermal and non-thermal food production are emerging trends focused on the future scope of food science and technology, and the technological and engineering challenges presented by this field. The adoption of innovative thermal and non-thermal food-processing technologies that retain the nutritional value of food products and increase their shelf life is likely to further drive the growth of the global market. Consumers are very concerned about the sustainability of the food products they are consuming.

Food Processing, which describes newly emerging technology for both academic and industrial research, is divided into two volumes (Advances in Thermal Technologies (Volume I), Advances in Non-Thermal Technologies (Volume II). Volume I covers the major aspects and emerging developments in thermal food-processing operations, and volume II covers advances in non-thermal food processing.

Thermal operations are the most common and conventional methods of obtaining and treating different products in food processing, the basic purpose of which is to reduce or destroy enzyme and microbial activity in food products. Volume I contains nine chapters that focus both on basic facts and on recent advances in thermal food-processing technologies.

Chapter 1 discusses various theoretical aspects, the basics of rate kinetics, and their application to the analysis of food quality indices. A number of practical issues related to food technology are also addressed. Chapter 2 describes innovative drying methods that have either already been incorporated into the agricultural processing environment or have enormous potential for application in the near future for food dehydration. Chapter 3 explains the evaporation method and the associated challenges and economic issues, as well as different types of commercially used evaporator.

Chapter 4 describes methods of industrial blanching, offering insights into modern, innovative blanching strategies and their effect on the quality and sensory characteristics of food. Chapter 5 discusses applications of thermal operations in food processing, illustrating the use of kinetic equations for the calculation of overall death time and lethality in various thermal operations. Chapter 6 describes the fundamental theory of deep-fat frying and how it is applied in practice, including the optimization of frying techniques and their reliance on the chemical reactions of oil, food, and other elements.

Chapter 7 demonstrates the fundamentals of crystallization, the key current areas of research in this field, and novel product development using revolutionary characterization and control methods. Chapter 8 describes the fundamentals of ohmic heating and its implementation in various food-processing operations including blanching, evaporation, dehydration, fermentation, sterilization, and pasteurization processes. Chapter 9 discusses various traditional and new extraction techniques, including pressurized liquid extraction, supercritical fluid extraction, microwave-assisted extraction, ultrasound-assisted extraction, pulsed electric field assisted extraction and high-pressure assisted extraction.

Kshirod Kumar Dash
Sourav Chakraborty

List of Figures

List of Tables

About the Editors

Kshirod Kumar Dash, PhD, is an Associate Professor and Head of the Department of Food Processing Technology, Ghani Khan Choudhury Institute of Engineering and Technology, Malda, West Bengal, India. His previous post was Assistant Professor in the Department of Food Engineering and Technology, Tezpur University, Tezpur, Assam. He obtained his B. Tech in Agricultural Engineering from the College of Agricultural Engineering and Technology, OUAT, Bhubaneswar, and his M. Tech in Dairy and Food Engineering, and PhD in Food Process Engineering from IIT Kharagpur, West Bengal, India. His post-doctoral research was carried out at Ohio State University, USA. He has published several research papers in peer-reviewed journals and participated in various national and international conferences. He has been involved in research projects funded by various funding agencies such as DST, MOFPI, and MSME. He is paper coordinator and content writer for Food Technology in the development of the e-course program of ePG Pathshala-Inflibnet. Dr Dash has received numerous awards and fellowships in recognition of his teaching and research achievements. His principal current research interests are drying technology, extraction, food process modeling, and heat and mass transfer in food processing.

Sourav Chakraborty, PhD, is currently Assistant Professor in the Department of Food Engineering and Technology, Tezpur University, Tezpur, Assam. He was previously a faculty member in the Department of Agricultural Engineering, Assam University, Silchar. He obtained his BTech in Agricultural Engineering from the North Eastern Regional Institute of Science and Technology (NERIST), and his MTech and PhD degrees in Food Engineering and Technology from Tezpur University. He has expertise in the food process engineering and food process modeling fields. Dr Chakraborty has published various research papers in national and international peer-reviewed journals. He has received honors at numerous international conferences for his groundbreaking research.

Contributors

Mohsin Bashir Aga
Department of Food Technology, Islamic University of Science and Technology, Awantipora, Kashmir, India

Aradhita Barmanray
Department of Food Technology, Guru Jambheshwar University Science and Technology, Hisar, Haryana, India

Sourav Chakraborty
Department of Food Engineering and Technology, Tezpur University, Tezpur, Assam, India

Aamir Hussain Dar
Department of Food Technology, Islamic University of Science and Technology, Awantipora, Kashmir, India

Ipsita Das
Centre for Technology Alternatives for Rural Areas, Indian Institute of Technology Bombay, Powai, Mumbai, India

Puja Das
Department of Agricultural Engineering, Triguna Sen School of Technology, Assam University, Silchar, India

Susanta Kumar Das
Department of Agriculture and Food Engineering, Indian Institute of Technology Kharagpur, West Bengal, India

Kshirod Kumar Dash
Department of Food Processing Technology, Ghani Khan Choudhury Institute of Engineering and Technology (GKCIET), Malda, West Bengal, India

Saptashish Deb
Department of Food Engineering and Technology, Sant Longowal Institute of Engineering and Technology, Longowal, Sangrur, Punjab, India

Pinki Deka
Department of Food Engineering and Technology, Tezpur University, Tezpur, Assam, India

Tabli Ghosh
Department of Food Engineering and Technology, Tezpur University, Tezpur, Assam, India

S.B. Kalse
Department of Post-Harvest Engineering, Killa-Roha, Raigad, Dr. Balasaheb Sawant Konkan Krishi Vidyapeeth, Killa-Roha, Raigad, Dapoli, India

Shafat Khan
Department of Food Technology, Islamic University of Science and Technology, Awantipora, Kashmir, India

C.B. Khobragade
Depatment of Agricultural Engineering, Triguna Sen School of Technology, Assam University, Silchar, India

Rahul Kumar
Department of Food Engineering and Technology, Tezpur University, Tezpur, Assam, India

Manish Kumar
Department of Food Technology, Guru Jambheshwar University Science and Technology, Hisar, Haryana, India

Ishrat Majid
Department of Food Technology, Islamic University of Science and Technology, Awantipora, Kashmir, India

Santanu Malakar
Department of Food Engineering, National Institute of Food Technology Entrepreneurship and Management, Sonepat, Haryana, India

Umexi Rani
Department of Food Engineering and Technology, Sant Longowal Institute of Engineering and Technology, Longowal, Sangrur, Punjab, India

Qurat Ul Eain Hyder Rizvi
Eternal University, Baru Sahib, Himachal Pradesh, India

Winny Routray
Department of Food Process Engineering, National Institute of Technology Rourkela, Rourkela, Odisha, India

Rafeeya Shams
Sher-e-Kashmir University of Agricultural Sciences and Technology, Jammu, India

Anurag Singh
National Institute of Food Technology and Entrepreneurship Management, Kundli, Haryana, India

Ajita Tiwari
Department of Agricultural Engineering, Triguna Sen School of Technology, Assam University, Silchar, India

1 Rate Kinetics in Thermal Food-Processing Operations

Kshirod Kumar Dash
Department of Food Process Technology, Ghani Khan Choudhury Institute of Engineering & Technology (GKCIET), Malda, West Bengal, India

Sourav Chakraborty
Department of Food Engineering and Technology, Tezpur University, Tezpur, Assam, India

Rahul Kumar
Department of Food Engineering and Technology, Tezpur University, Tezpur, Assam, India

Tabli Ghosh
Department of Food Engineering and Technology, Tezpur University, Tezpur, Assam, India

CONTENTS

1.1 INTRODUCTION

Thermal sterilization, pasteurization, and blanching techniques are used extensively to control bacteria, fungi, and other microorganisms, as well as undesirable enzymes, in foods. Thermal processing, sometimes done in order to meet food safety and stable shelf-life requirements, involves high-temperature heating for a long time at a high temperature, to promote physical and chemical reactions that preserve food quality (Fellows, 2009). The major quality parameters affected by thermal treatment are changes in color, unusual flavors, and loss of freshness or nutrients during and after thermal processing (Liaotrakoon et al., 2013; Song et al., 2003; Wen et al., 2010; Zabbia et al., 2011). This is especially the case for prepackaged solid or semisolid foods where slow heat transfer occurs. Conventional surface heating methods face a challenge to raise the temperature of the central core of the food.

Novel thermal processing techniques play a vital role in improving heating efficiency and minimizing loss of quality. In order to develop better quality food, electro-heating techniques, such as ohmic (OH), microwave (MW), and radio frequency (RF) heating are mostly used in place of conventional heating methods (Liu et al., 2011; Pereira et al., 2008; Wang et al., 2012). In Europe, microwaves at frequencies of 846 or 2450 MHz are used in packaged food production (Tang et al., 2002). Volumetric heating, in which electrical energy is directly converted into heat, decreases heating time and increases heating rate, helping to avoid the loss of quality associated with slow heating methods (Marra et al., 2009).

The design of thermal process systems requires a comprehensive understanding of the thermal properties and quality attributes of foods, and the quantitative changes in target microorganisms and enzymes (Stoforos, 1995; Holdsworth and Simpson, 2008; Simpson, 2010). Appropriate thermal process conditions need to be selected in order to control microorganisms and enzymes, and minimize degradation of food quality. In the literature, various kinetic studies on thermal inactivation of microorganisms and enzymes have been reported (Van Boekel, 1996; Dolan, 2003; Hindra and Baik, 2006; Barsa et al., 2012; Bermudez-Aguirre and Corradini, 2012).

1.2 THE THEORETICAL BASIS OF REACTION RATE KINETICS IN THERMAL PROCESSING

During thermal processing numerous changes occur in food material. The eventual loss of quality is affected by several interactions and complex reactions rather than a single elementary step. Studying simple food systems rather than real foods is an effective first step in kinetic modeling (Wedzicha et al., 1993). To quantify individual attributes of an ideal food system, chemical reaction kinetics can be expressed as a general rate law (Van Boekel, 1996; Steinfeld et al., 1998; FDA, 2000).

$$\frac{dA}{dt} = \pm kA^{n} \tag{1.1}$$

Here, k stands for the rate constant (1/min), t stands for the reaction time (min), n stands for the reaction order, and A stands for a quantitative value for a quality attribute, enzyme activity, or microorganism population.

The essence of the kinetic study of food quality change is the quantification of a quality attribute as a function of heating time at a certain temperature. The quantification is done by using temperature-dependent reaction rate constants and determining the reaction order. The order of kinetics is determined by fitting the experimental data to a preselected reaction order model based on best fit. A number of kinetic models are used to describe food quality changes.

1.2.1 Zero-Order Reaction

By putting n = 0 in Equation (1.1), the mathematical expression can be written as follows:

$$\frac{dA}{dt} = -kA^{0} = k \tag{1.2}$$

Integrating Equation (1.2) with limits A = 0 to A and t = 0 to t, the final equation can be written as follows:

$$\int_{A_0}^{A} dA = -k\int_{0}^{t} dt \tag{1.3}$$

$$A = A_0 - kt \tag{1.4}$$

The half-life of a zero-order reaction is defined as the time required for the reduction of half of the concentration. The mathematical equation for the half-life is given as follows:

$$t_{\frac{1}{2}} = \frac{A_0}{2k} \tag{1.5}$$

1.2.2 First-Order Reaction

By putting n = 1 in Equation (1.1), the mathematical expression can be written as follows:

$$\frac{dA}{dt} = -kA \tag{1.6}$$

Integrating Equation (1.6) with limits A = 0 to A and t = 0 to t, the final equation can be written as follows:

$$\int_{A_0}^{A} \frac{dA}{A} = -k\int_{0}^{t} dt \tag{1.7}$$

$$\ln\frac{A}{A_0} = -kt \tag{1.8}$$

Multiplying each side of the equation with the exponent (exp), the final expression can be written as follows:

$$A = A_0 \exp(-kt) \tag{1.9}$$

The mathematical expression for the half-life of the first-order reaction is given as follows:

$$t_{\frac{1}{2}} = \frac{0.693}{k} \tag{1.10}$$

1.2.3 Second-Order Reaction

By putting n = 2 in Equation (1.1), the mathematical expression can be written as follows:

$$\frac{dA}{dt} = -kA^2 \tag{1.11}$$

Integrating Equation (1.11) with limits A = 0 to A and t = 0 to t, the final equation can be written as follows:

$$\int_{A_0}^{A} \frac{dA}{A^2} = -k\int_{0}^{t} dt \tag{1.12}$$

$$\frac{1}{A} = kt + \frac{1}{A_0} \tag{1.13}$$

The mathematical expression for the half-life of the second-order reaction is given as follows:

$$t_{\frac{1}{2}} = \frac{1}{kA_0} \tag{1.14}$$

Here, A_0 stands for initial value of food quality attributes at t = 0, k is reaction rate constant, $t_{\frac{1}{2}}$ stands for half-life.

1.2.4 Pseudo-First-Order Model

The pseudo-first-order kinetic model can be expressed mathematically as follows:

$$\log(q_e - q_t) = \log(q_e) - \frac{k}{2.303}t \quad (1.15)$$

1.2.5 Pseudo-Second-Order Model

The pseudo-second-order kinetic model can be expressed mathematically as follows:

$$\frac{t}{q_t} = \frac{1}{kq_e^2} + \left(\frac{1}{q_e}\right)t \quad (1.16)$$

Here, q_t and q_e are values of a food quality attribute at time t and equilibrium respectively, k is reaction rate constant and t is treatment or heating time.

1.2.6 Special Kinetic Models

For thermal processing of food, generally zero-, first-, or second-order kinetic models are employed. Hiwilepo-van Hal et al. (2012) reported the inadequacy of a simple first-order model for describing the thermal degradation of vitamin C. Ko et al. (2007) also reported biphasic behaviors in the kinetic modeling of texture change during thermal processing. Several alternative models have been proposed to improve description of thermal degradation, and these are described below.

1.2.6.1 Two Fractions or Biphasic Model

Liing and Lund (1978) reported the biphasic model for describing the inactivation thermal kinetics of an enzyme system. The mathematical expression of the model is given as follows:

$$A = A_s \exp(-k_s t) + A_l \exp(-k_l t) \quad (1.17)$$

Here, A_s and A_l stand for activities of the stable and the labile fractions and k_s and k_l refer to inactivation rate constant of stable and labile fractions.

1.2.6.2 Weibull Model

The Weibull model was originally developed to describe the failure of a given system subjected to stress conditions over time. The mathematical expression of the model is given as follows (Van Boekel, 2002):

$$\log\frac{A}{A_0} = -bt^n \quad (1.18)$$

Here, A is the residual activity and A is the initial activity, and b and n are the scale and shape factors, respectively. For n < 1, upper concavity is observed for the isothermal semi-logarithmic curve and for n > 1 downward concavity is observed. A first-order kinetic model can be obtained for n = 1, showing the greater flexibility of the Weibull model for biological systems.

1.2.6.3 Fractional Conversion Model

This model, which measures the extent of a reaction as an indication of quality change, is mostly used in chemical engineering, and is given mathematically as follows:

$$A_t = A_\infty + (A_0 - A_\infty)\exp(-kt) \tag{1.19}$$

Here, A_0 refers to initial activity, A_t is activity after a certain treatment time (t), A_∞ is residual activity after prolonged treatment time, t is treatment time, and k stands for reaction rate.

When the reaction is completed, A_∞ for an irreversible reaction would be zero. Hence, Equation (1.19) can be modified as follows:

$$A_t = A_0 \exp(-kt) \tag{1.20}$$

However, for some quality indices, such as the concentration of bioactive compounds, antioxidant capacity, and color of foods after high-temperature treatments, the value of A_∞ may be approximately zero. But in the case of texture properties enhanced due to the effect of thermal treatment, A_∞ may also have non-zero characteristic values. A fractional conversion model can be effectively used for the characterization of textural degradation during thermal processing (Peng et al., 2014).

1.2.7 Temperature Dependence of Reaction Rate Constant

The Arrhenius equation is normally used to explore the dependence of quality parameters on temperature. The relationship between the reaction rate constant and temperature can be given as follows:

$$k = k_0 \exp\left(-\frac{E_a}{RT}\right) \tag{1.21}$$

Here, k_0 stands for the rate constant, R stands for ideal gas constant (8.314 J/mol·K), and T refers to the absolute temperature (K). E_a is the activation energy (J/mol). The minimum energy required to start a chemical reaction is known as activation energy. When an appreciable number of molecules with energy equal to or greater than the activation energy interact, a chemical reaction takes place at a reasonable rate. With the increase in temperature, the number of molecules with energy higher than the activation energy increases, improving the rate of reaction. Rate to temperature is therefore an important parameter indicating the sensitivity of the reaction.

An alternative method of estimating kinetic parameters for the Arrhenius relationship is also recommended (Theodore et al., 1997; Peleg et al., 2012).

$$k = k_{ref} e^{-\frac{E_a}{R}\left(\frac{1}{T}-\frac{1}{T_{ref}}\right)} \tag{1.22}$$

Here, T_{ref} is the reference temperature (K) corresponding to the mean of the temperature range, and k_{ref} is the reaction rate at the reference temperature.

The Q10-value, also an important concept for describing the rate of quality changes with temperature, is defined as follows (Lund, 1977; Fu and Labuza, 1993):

$$k = \frac{k\left(T + 10^{\circ}C\right)}{k_T} \tag{1.23}$$

Here T is temperature in °C.

1.2.8 Q10-Value

Assuming the Arrhenius relationship, the Q10-value can be related to the activation energy by the following equation (Buransompob et al., 2003):

$$\log Q_{10} = \frac{4.34E_a}{RT\left(T + 10^{\circ}C\right)} \tag{1.24}$$

The Q10-value is commonly used to predict storage time at lower temperatures, based on shelf-life tests conducted at higher temperatures. Due to its strong dependence on temperature, the Q10-value is applied across a small temperature range of 10–20 °C (Labuza, 1984). The Q10-value is required to design accelerated shelf-life tests for thermally processed products, such as walnuts, almonds, powdered gabiroba pulp, freshly cut pineapple slices, and low-acid foods (Yoon et al., 2009).

1.2.9 Z-Value

For experimental data outside the temperature limits, translation of factors (E_a- and z-values) from one model to the other can, however, lead to discrepancies. Where the Arrhenius relationship holds, z-values can also be expressed by the following mathematical relationship within a given temperature range $T_1 \sim T_2$ $(T_1 < T_2)$:

$$z = \frac{2.303RT_1T_2}{E_a} \tag{1.25}$$

The following equation describes the rate constant using both the Arrhenius equation and the z-value model (Saguy et al., 1978):

$$z = \frac{2.303RT^2}{E_a} \tag{1.26}$$

Here, T refers to a temperature corresponding to the mean over the temperature range $T_1 \sim T_2$ in experiments.

The relationship between the z-value and Q10-value can be expressed as follows:

$$z = \frac{10}{\log Q_{10}} \tag{1.27}$$

Each parameter – D-value, z-value, E_a, k, Q10, and T – can easily be converted if two of the model parameters are known. There are both correlations and differences between the Arrhenius equation and the z-value model. Within a given temperature range they are nearly the same, but the z-value model gives a higher predicted k-value than the Arrhenius equation (Fujikawa and Itoh, 1998; Jonsson et al., 1977; Saguy et al., 1978).

1.3 METHODS FOR ESTIMATING KINETIC PARAMETERS IN THERMAL PROCESSES

In order to estimate changes in the quality of food materials, a given temperature–time heating process history is generally applied to kinetic models (Dolan, 2003). Preliminary tests are conducted on the kinetic models being developed to select test conditions over the maximum timespan at any given temperature, in order to ensure that the required extent and range of quality changes can be observed. As a precautionary measure, the effect of the time taken for the sample to reach the set elevated temperature should also be excluded (Chung et al., 2007).

For kinetic studies, two experimental methods are widely used: direct and indirect heating. In direct heating, prepackaged samples are directly immersed in the heating medium, such as a water bath. Although this method works well for conventional treatments, it is not suitable for kinetic studies of large samples due to the non-uniform temperature distribution between the food surface and the center. It also fails to provide isothermal conditions, due to the longer time required for whole prepackaged samples to reach the target temperature. The kinetic parameters determined by this method only work well for the treatment of food materials under the same heating conditions.

Indirect heating methods use a small sample volume or thin product to achieve suitable isothermal conditions. Containers suitable for indirect heating, according to the type of food product, are cans, glass beakers, glass tubes, screw-cap tests, Pyrex glass vials, hermetically sealed metal cells or tubes, and capillary tubes (Ling et al., 2015).

1.4 APPLICATION OF KINETIC MODELS IN THERMAL PROCESSING OF FOOD

Appearance, flavor, texture, and nutrition are important quality attributes in foods (Bourne, 2002). This section discusses the thermal kinetics of these attributes, including color, bioactive compounds, and texture.

1.4.1 Color Kinetics in Food Processing

1.4.1.1 Theoretical Considerations

The kinetics of color change in food products are commonly studied using zero- and first-order reaction and fractional conversion models (Onwude et al., 2016).

$$\frac{dC}{dt} = -k(C)^n \tag{1.28}$$

where C represents the concentration of colorimetric parameters of specific food products with respect to time t; dC/dt represents the rate of change of the three colorimetric concentrations over time t; k represents the rate constant; and n represents the order of the reaction (zero-order reaction, n = 0; first-order reaction, n = 1). For a zero-order reaction, n = 0, and the colorimetric parameters are independent of reaction rate:

$$\frac{dC}{dt} = -k \tag{1.29}$$

The integration of Equation (1.29) is represented in Equation (1.30), where concentration C changes from C_O (t = 0) to C (t = t).

$$C = C_0 - kt \tag{1.30}$$

In this regard, Equation (1.30) represents the zero-order reaction rate model.

Additionally, for first-order reaction rates, the reaction rates are dependent on the color coordinates as represented in Equation (1.31), and the integration of Equation (1.31) is represented in Equation (1.32).

$$\frac{dC}{dt} = -k(C) \tag{1.31}$$

$$C = C_0 e^{-kt} \tag{1.32}$$

The first-order reaction kinetic model for studying the color coordinates in food products can be estimated using fractional conversion, as shown in Equation (1.33).

$$\frac{C - C_f}{C_0 - C_f} = e^{-kt} \tag{1.33}$$

where C_0, C and C_f are the values of sample color at t = 0, any time, and final color values, respectively.

Additionally, the rate constant follows the Arrhenius relationship, which is temperature dependent, and the relationship between rate constants with color parameters and drying temperatures can be determined using the following equation:

$$K = K_0 \exp\left[-\frac{E_a}{R}\left(\frac{1}{T + 273.15} - \frac{1}{T_{ref} + 273.15}\right)\right] \tag{1.34}$$

where E_a denotes the activation energy, kJ/mol; R is the universal gas constant, 8.31451×10^{-3} kJmol^{-1}k^{-1}; k_0 is pre-exponential factor; and T_{ref} and T are the reference and air temperatures, °C.

1.4.1.2 Application of Color Kinetics

Drying is widely used in thermal processing of food products to extend shelf life. The drying of food products also helps to reduce volume, which is a further advantage for transportation. Drying affects color and quality if processing is of longer duration. Mathematical modeling of color-based reaction kinetics, while complex, helps to determine parameters such as order of reaction kinetics, rate constant, and activation energy, and to optimize processing conditions. The quality of raw materials and physico-chemical changes, such as physical and chemical changes during processing and storage, also affect the quality of dehydrated food products (Nadia et al., 2011).

Color kinetics in food materials are commonly observed to follow zero-order reaction kinetics. However, different phenomena such as total color difference, carotenoid content, and non-enzymatic browning, can follow several different reaction kinetics (see Table 1.1). The color change kinetics of several food products during processing are discussed below.

The color of grape juice and grape leather is an important factor for customer acceptability, as color is directly related to flavor. In a study of changes in color parameters during boiling, cooking and pulverizing of grape juice (Maskan et al., 2002), changes in color parameters were observed during boiling, whereas the color parameter "a" changed during hot-air dying. The reaction associated with color change is observed to follow zero-order reaction kinetics and is temperature dependent, following the Arrhenius equation. Juice concentration processing changes the color parameters: the L-value decreases, whereas a- and b-values increase, which may be due to anthocyanin degradation and Maillard reaction. The color of asparagus is sensitive to various thermal treatments including cooking and blanching, and asparagus may appear olive green in color after prolonged cooking. Thermal treatment (between 70 and 98 °C) of green asparagus causes color changes that follow first-order reaction kinetics (Lau et al., 2000). Kinetic total carotenoid degradation and $L \times b$ values during hot-air drying (temperatures: 50, 60 and 70 °C) of jackfruit bulb slices follow first-order reaction kinetics; however, total color difference and non-enzymatic browning can be studied using zero-order reaction kinetics (Saxena et al., 2012). Total carotenoid degradation, $L \times b$ values, total color difference, and non-enzymatic browning reaction kinetics during hot-air drying of jackfruit bulb slices have activation energies of 27.21, 29.96, 55.61, and 27.48 kJ/mol, respectively. The higher the activation energy, the greater the sensitivity of color to temperature. Interestingly, hot-air, microwave, and hot-air/microwave-finish drying affect the color of kiwi fruits, with zero- and first-order kinetic models describing color changes during drying. In a study by Maskan (2001), the color parameter L- and b-values are properly described by both zero- and first-order reaction kinetics, whereas a-values followed first-order reaction kinetics. The modeling of color parameters to detail browning index values during drying of oyster mushrooms (temperatures: 45, 55, and 65 °C) follows both zero-order and first-order reaction kinetics (Engin, 2019). On the other hand, total color change values follow zero-order kinetics, where the reaction rate constants are 0.058, 0.065, and 0.089 for drying temperatures 45, 55, and 65 °C, respectively. The drying of okra at selected microwave powers significantly affects characteristics such as color, hardness, sensory attributes, and moisture diffusivity (Aamir and Boonsupthip, 2017). In this microwave drying process, the

TABLE 1.1
Reaction Kinetics Parameters in Drying of Various Food Products

S/N	Types of Food Product	Types of Kinetic Model	Reaction Kinetic Parameters Color Parameters	References
1.	Grape juice concentration	Zero-order reaction kinetics	L value, C_0 = 64.96, k (min^{-1}) = 0.178, r = 0.99	Maskan et al. (2002)
2.	Grape juice concentration	Zero-order reaction kinetics	a* values, C_0 = – 3.33, k (min^{-1}) = 0.100, r = 0.988	Maskan et al. (2002)
3.	Grape juice concentration	Zero-order reaction kinetics	b* values, C_0 = 18.3, k (min^{-1}) = 0.90, r = 0.997	Maskan et al. (2002)
4.	Asparagus (Heat treatment between 70 and 98 °C)	First-order reaction	E_a: 13.1 ± 0.2 kcal/mol, k: 0.0066 ± 0.0002 min^{-1} (84 °C)	Lau et al. (2000)
5.	Drying Jackfruit bulbs (50, 60, 70 °C)	First-order reaction kinetics	k = 0.069 h^{-1} (50 °C), 0.101 (60 °C), 0.133 (70 °C), E_a = 29.96 kJ/mol	Saxena et al. (2012)
6.	Drying Jackfruit bulbs	Zero-order reaction kinetics	k = 0.553 h^{-1} (50 °C), 0.846 h^{-1} (60 °C), 0.995 h^{-1} (70 °C), E_a = 27.21 kJ/mol	Saxena et al. (2012)
7.	Hot air and microwave drying of kiwifruits	Zero-order reaction kinetics	k_0 = 0.0667 ± 0.0288 min^{-1}, C_0 = – 1.314 ± 3.22	Maskan (2001)
8.	Hot air and microwave drying of kiwifruits	Zero-order reaction kinetics	k_1 = 0.2452 ± 0.0757 min^{-1}, C_0 = – 0.000 ± 0.00	Maskan (2001)
9.	Microwave drying of okra	Zero-order reaction kinetics	k = 2.96 min^{-1} (440 W), 2.86 min^{-1} (618 W), 3.66 min^{-1} (800 W) E_a: 4.44 kJ/mol	Aamir and Boonsupthip (2017)
10.	Microwave drying of okra	Zero-order reaction kinetics	k = 0.66 min^{-1} (440 W), 1.40 min^{-1} (618 W), 1.41 min^{-1} (800 W), E_a: 19.97 kJ/mol	Aamir and Boonsupthip (2017)
11.	Microwave drying of okra	First-order reaction kinetics	k = 0.27 min^{-1} (440 W), 0.21 min^{-1} (618 W), 0.24 min^{-1} (800 W), E_a: 3.72 kJ/mol	Aamir and Boonsupthip (2017)
12.	Microwave drying of okra	First-order reaction kinetics	k = 0.23 min^{-1} (440 W), 0.26 min^{-1} (618 W), 0.29 min^{-1} (800 W) E_a: 5.47 kJ/mol	Aamir and Boonsupthip (2017)
13.	Microwave drying of basil	Zero-order reaction kinetics	k = 0.1516 min^{-1}; C_0 = 35.430 (180 W), k = 0.3255 min^{-1}; C_0 = 35.274 (360 W), k = 0.5047 min^{-1}; C_0 = 35.092 (540 W)	Demirhan and Özbek (2009)
14.	Microwave drying of basil	First-order reaction kinetics	k = 0.0098 min^{-1}; C_0 = 14.538 (180 W), k = 0.0191 min^{-1}; C_0 = 14.313 (360 W), k = 0.0305 min^{-1}; C_0 = 13.684 (540 W)	Demirhan and Özbek (2009)

color parameters L- and a-values are well described by the zero-order kinetic model, whereas b and total color difference are described by first-order models (see Table 1.1). Similarly, color changes during microwave drying of dried basil are affected by microwave power and sample size (Demirhan and Özbek, 2009). The color parameters a- and b-values follow the first-order kinetic model, whereas the L-values and total color change follow the zero-order reaction. The values of color parameter a for grapes dipped in potassium carbonate and ethyl oleate solutions (dipping time 1, 2, and 3 min at 30, 40, 50, and 60 °C) follow zero-order reaction kinetics (Bingol et al., 2012). Interestingly, the carotenoid content of food products is greatly affected by thermal processes, including drying and heating, due to processing conditions such as oxygen concentration, light, and heat. To optimize processing conditions, color change kinetics need to be studied to determine reaction order, rate constant, and activation energy. The activation energies during hot-air drying (50, 60 and 70 °C) of jackfruit bulb slices are 29.96, 27.21, 27.48, and 55.61 kJ/mol for L × b values, total color difference, non-enzymatic browning, and carotenoid content, respectively (Saxena et al., 2012).

Microbial safety in food product storage can be maintained via several food-processing techniques. Infrared heating can improve the microbial safety of dry roasted almonds using different methods such as traditional hot-air roasting, IR roasting, and sequential infrared and hot-air roasting (Yang et al., 2010). Of the three methods, sequential infrared and hot-air roasting is the most promising for producing roasted pasteurized almonds; the total color change follows zero-order reaction kinetics with activation energy 67.60 kJ/mol. The reaction kinetics for texture and color changes in rehydrated fig fruits are based on processing temperature and time, with textural changes well described by first-order reaction kinetics (Ansari et al., 2015). The temperature dependency of the color and textural factors can also be described by the Arrhenius model. However, the temperature dependency of some food products can also be described using a linear function relationship (Kahyaoglu and Kaya, 2006).

Fruits and vegetables, spices and various plant materials are excellent sources of bioactive components, providing several health benefits as well as environmental sensitivity. Other thermal processing techniques for food products, such as blanching, frying, sterilization, and pasteurization, need to be optimized to improve antioxidant, color, textural properties, total carotenoid content, and so on. Several parameters influence product quality in frying, including thickness of sample, oil temperature, oil type, sample type, sample properties, and so on. The frying of tofu significantly changes its physico-chemical and color properties: the yellowness factor of tofu increased from 18.8 to 34.5 during deep-fat frying, exponentially decreasing to 29.4 (Baik and Mittal, 2003). Deep-fat fried tofu (temperature 147–172 °C for 0–5 min) is a popular product, and frying conditions are critical for obtaining optimum viscoelasticity (Baik and Mittal, 2003). Viscoelastic properties can be characterized using the stress relaxation test, and can be expressed with two Maxwell elements and a residual spring in parallel, where the elastic modulus and relaxation time follow zero-order reaction kinetics. The viscoelastic behavior of food materials is generally studied using dynamic mechanical analysis, and activation energies range from 1.00×10^5 to 1.49×10^5 J/mol. Thus several food product processing conditions need to be optimized to obtain a good texture.

1.4.2 Kinetics of Bioactive Compounds

Thermal treatments applied in food processing to inactivate microorganisms and enzymes affects the quality of various bioactive compounds. Study of degradation kinetics is therefore necessary in order to control these compounds. The first-order kinetic model is used to study color degradation kinetics of these compounds. Though classical first-order models fit some data well, the degradation kinetics of thiamin for high-temperature treatment can be better predicted using the second-order reaction model (Kong et al., 2007). The thermal degradation curves of the lutein of blood orange juice over a temperature range of 45–90 °C fit the second-order reaction model (Hadjal et al., 2013). Harbourne et al. (2008) reported thermal degradation kinetics of blackcurrant anthocyanins in a model juice system over 4–140 °C by applying the pseudo-first-order reaction kinetic model. Zanoni et al. (2003) reported the application of pseudo-zero-order reaction kinetics for the antioxidant activity of hydrophilic extract during sterilization. The biphasic first-order and fractional conversion models are also suitable for the thermal degradation of health-related functional compounds (Hiwilepo-van Hal et al., 2012; Jaiswal et al., 2012; Vieira et al., 2000). Table 1.2 shows reaction kinetics models for the bioactive compounds of various food products.

TABLE 1.2
Reaction Kinetics Models for the Bioactive Compounds of Various Food Products

Type of Bioactive Component	Type of Product	Types of Heating Methods	Type of Kinetic Model	Details of Kinetic Parameters (k value in min^{-1})	References
Vitamin C	Cupuacu nectar	Water-bath heating	Fractional first order	3.2×10^{-2}	Ling et al. (2015)
			First order	1.3×10^{-2}	
	Orange juice	Ohmic heating	First order	1.6×10^{-1}	
		Microwave heating	First order	5.0×10^{-2}	
Thiamin	Pink salmon	Oil-bath heating	Second order	0.1	
Carotenoids	Citrus juice	Oil-bath heating	First order	5.3×10^{-3}	
Anthocyanine	Blackcurrant Juice	Water-bath heating	Pseudo first order	3.5×10^{-4}	
	Raspberries	Oil-bath heating	First order	2.0×10^{-2}	
Total Phenolics	Cabbage	Water-bath heating	Fractional first order	4.2×10^{-1}	
Antioxidant Activity	Fruit extract of Mulberry	Water-bath heating	Zero order	2.2×10^{-2}	

1.4.3 Texture Kinetics

Texture is very important for evaluating finished product quality and consumer acceptability. Texture profile analysis is normally used to evaluate different textural properties of food products. Hardness, springiness, and firmness are most often studied to quantify the effect of thermal degradation on textural kinetics (Jaiswal et al., 2012; Lau et al., 2000; Nisha et al., 2006). Ovissipour et al. (2013) reported that zero-order kinetics models adequately predicted compression force of whole mussel during thermal pasteurization. Kong et al. (2007) and Brookmire et al. (2013) reported the first-order kinetic model as predicting sheer force of salmon fillets during thermal processing. Other kinetic models such as the biphasic first-order model, fractional conversion model, and Weibull-log-logistic model also adequately predict the thermal degradation of textural properties. Table 1.3 gives details of the reaction kinetics models for the texture properties of various food products.

TABLE 1.3
Reaction Kinetics Models for the Texture Properties of Various Food Products

Types of Product	Types of Heating Methods	Types of Texture Parameters	Type of Kinetic Model	Details of Kinetic Parameters (k value in min^{-1})	References
Green asparagus	Water-bath heating	Shear stress	First order	1.6×10^{-2} (for butt portion) 2.3×10^{-2} (for middle portion) 2.7×10^{-2} (for budd)	Ling et al. (2015)
Carrots	Water-bath and oil-bath heating	Hardness	Fractional first order	1.5×10^{-1}	
Pink Salmon	Oil-bath heating	Shear force	First order	1.9×10^{-2}	
Whole blue mussel	Oil-bath heating	Compression force	Zero order	2.6×10^{-2}	
Pumpkin	Water-bath heating	Firmness	Fractional first order	3.9×10^{-1}	
Mushroom	Water-bath heating	Hardness	Biphasic first order	2.7×10^{-1}	
Litchi	Water-bath heating	Firmness	First order Fractional first order Weibull	5.4×10^{-3} 1.5×10^{-2} 2.7×10^{-2}	
Cabbage	Water-bath heating	Firmness	First order	9×10^{-2}	

1.5 NUMERICAL PROBLEMS RELATED TO THERMAL KINETICS OF FOOD MATERIALS

Example 1:

The firmness values of a protein gel as a function of time of heating are shown in Table 1.4 (Toledo et al., 2007). Apply firmness data to the fractional conversion model and determine the value of the reaction rate constant.

Solution:

From Equation (1.19), we know that $A_t = A_\infty + (A_0 - A_\infty)\exp(-kt)$.
Here, $A_0 = 0$
So the equation can be written as follows:

$$\frac{A_\infty - A_t}{A_\infty} = \exp(-kt)$$

$$\rightarrow \ln\left(\frac{A_\infty - A_t}{A_\infty}\right) = -kt$$

$$\text{Suppose}\ \frac{A_\infty - A_t}{A_\infty} = P$$

$$\text{So,}\ln(P) = -kt \tag{1.35}$$

The values of P at each time, t, are given in Table 1.5. By plotting (Figure 1.1) P in the y-axis and t in the x-axis in Microsoft Excel, the slope is determined (intercept = 0) as –0.763. Based on Equation (1.35), the reaction rate (k) is found to be 0.763 min^{-1}.

TABLE 1.4
Firmness Value of a Protein at Different Heating Times

Time (min)	0	1	2	3	4	5	10	20
Firmness (g)	0	6.01	8.41	9.36	9.75	9.9	10.2	10.2

TABLE 1.5
Results of Fractional Conversion Model Fitting

Time (min)	Firmness (g)	P
0	0	0.00
1	6.01	–0.89
2	8.41	–1.74
3	9.36	–2.50
4	9.75	–3.12
5	9.9	–3.53

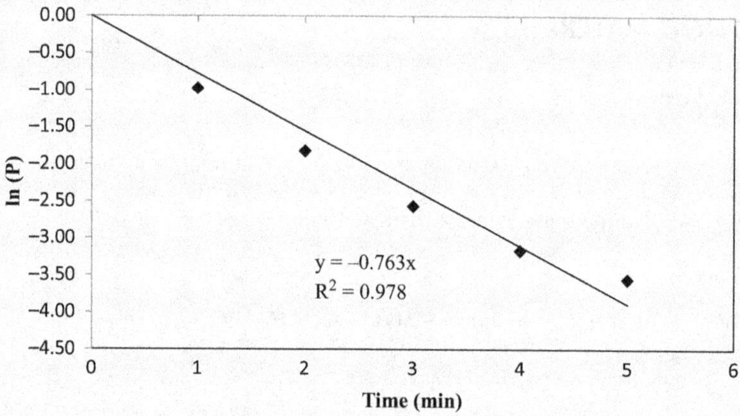

FIGURE 1.1 Estimation of k using Equation 1.19

Example 2:

Vikram et al. (2005) reported thermal degradation kinetics of vitamin C in orange juice during conventional thermal treatment. They fitted first-order kinetics with the data obtained at various temperatures (Table 1.6). Calculate (a) D-value for each temperature and (b) activation energy (E_a).

Solution:

(a) We know that

$$D = \frac{2.303}{k}$$

So, based on this relationship, D-values are calculated for each temperature. The results are shown in Table 1.7.

(b) It is known that $k = k_0 \exp\left(-\frac{E_a}{RT}\right)$

So, by applying ln (natural log) on both sides of the equation, the final expression can be written as follows:

$$\ln(k) = \ln(k_0) - \frac{E_a}{RT}$$

By plotting ln (k) in y-axis and 1/T in x-axis (Figure 1.2), the values of k_0 and E_a are determined. The slope is found to be −4793.

Hence,

$$-\frac{E_a}{R} = -4793$$

The value of E_a = 39.84 kJ/mol K (by taking R = 8.314 J/mol).

TABLE 1.6
k-Value Obtained for the Vitamin C Data of Orange Juice

Temperature (°C)	k value (min^{-1})
50	0.0351
60	0.0462
75	0.0852
90	0.1784

TABLE 1.7
D-values Calculated for Each Temperature

Temperature (°C)	k value (min^{-1})	D value (min)
50	0.0351	66
60	0.0462	50
75	0.0852	27
90	0.1784	13

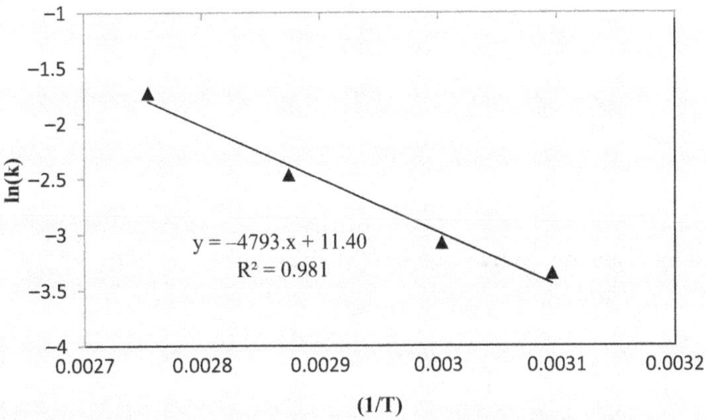

FIGURE 1.2 Estimation of E_a using Equation 1.21

Some Numerical Problems

Q1. Suppose second-order kinetics is followed for the conversion of molecule X to Y. If the concentration of X is increased by a factor of three, what will be the effect on the rate of formation of Y?

Solution:

Here the reaction is:

$$X => Y \text{ and } r = K[X]^2$$

If the concentration of X is increased by a factor of three, let the new rate be r′

$$r' = K[3X]^2 = K[X]^2 \times 9 = 9r$$

Hence the rate of formation of Y will increase by a factor of 9.

Q2. For a first-order reaction the half-life of a is 20 min. To accomplish 90% completion, how long will the reaction take?

Solution:

$$t_{1/2} = 20\,\text{min} => \frac{0.693}{K_1} = 20 \qquad \text{(a)}$$

For 90% completion, if a = 100, x = 90

$$K_1 = \frac{2.303}{t_{90\%}} \times \log\frac{100}{10}$$

$$= \frac{2.303}{t_{90\%}}$$

$$\frac{2.303}{K_1} = t_{90\%} \qquad \text{(b)}$$

Dividing Equation (a) by Equation (b) gives:

$$\frac{0.693}{2.303} = \frac{20}{t_{90\%}}$$

$$=> t_{90\%} = 66.46\,\text{min}.$$

Q3. The experimental data for the reaction $2A + B_2 \rightarrow 2AB$ are given in Table 1.8. Formulate the equation for the rate of reaction.

Solution:

From Table 1.8, we observe that when the concentration of B_2 is doubled, the rate is doubled. Hence the order of reaction with respect to B_2 is one. When concentration of A is doubled, the rate remains unaltered. So the order of reaction with respect to A is zero.

TABLE 1.8
Experimental Data for the Reaction $2A + B_2 \rightarrow 2AB$

Expt. No.	[A] (mol L−1)	$[B_2]$ mol L^{-1})	Rate (mol L^{-1} s^{-1})
1	0.50	0.50	1.6×10^{-4}
2	0.50	1.00	3.2×10^{-4}
3	1.00	1.00	3.2×10^{-4}

Therefore,

$$\text{Rate} = k[B_2]^{\alpha}[A]^{0} = k[B_2]^{\alpha}$$

Thus rate equations for a different concentration and the rate of B can be written as follows:

$$1.6\times10^{-4} = k[0.5]^{\alpha} \text{ and } 3.2\times10^{-4} = k[1]^{\alpha}.$$

By division the value of α is found to be 1.
Hence, the final rate equation will be as follows:

$$\text{Rate} = k[B_2]$$

Q4. When log k was plotted against 1/T, for a first-order reaction a straight line with a slope of −6000 was obtained. Determine the activation energy of the reaction.

Solution:

$$\text{Slope} = \frac{-E_a}{2.303R} = -6000$$

$$E_a = 6000\times2.303\times8.314$$

$$E_a = 1.148\times10^{5}\ \text{J}$$

$$E_a = 27483.935\,\text{Cal} = 27.48\,\text{Kcal}$$

1.6 CONCLUSION

- This chapter has discussed different kinetic models and their application to controlling the thermal processing of foods.
- Thermal treatments are extensively used in the food industry to control pathogenic and spoilage microorganisms and enzymes.
- The major concern during thermal processing is degradation of food, adversely affecting its quality.
- Kinetic study of food quality changes is essential for the design of appropriate thermal treatments. This chapter mainly focuses on the kinetic study of thermal treatments to inactivate microorganisms and enzymes in food.
- The chapter discussed the theoretical basis for kinetic study, common and special kinetic models related to major quality attributes of food, and the application of kinetic models to the development of thermal treatments.

REFERENCES

Aamir, M. and W. Boonsupthip. 2017. Effect of microwave drying on quality kinetics of okra. *J. Food Sci. Tech.* 54(5): 1239–1247.

Ansari, S., N. Maftoon-Azad, E. Hosseini, A. Farahnaky and G.H. Asadi. 2015. Kinetic of color and texture changes in rehydrated fig. *J. Agri. Sci.* 21(1): 108–122.

Baik, O.D. and G.S. Mittal. 2003. Kinetics of tofu color changes during deep-fat frying. *LWT-Food Sci. Tech.* 36(1): 43–48.

Barsa, C.S., M.D. Normand and M. Peleg. 2012. On models of the temperature effect on the rate of chemical reactions and biological processes in foods. *Food Eng. Review.* 4(4): 191–202.

Bermudez-Aguirre, D. and M.G. Corradini. 2012. Inactivation kinetics of Salmonella spp. under thermal and emerging treatments: A review. *Food Res. Inter.* 45(2): 700–712.

Bingol, G., J.S. Roberts, M.O. Balaban and Y.O. Devres. 2012. Effect of dipping temperature and dipping time on drying rate and color change of grapes. *Dry. Tech.* 30(6): 597–606.

Bourne, M. 2002. *Food Texture and Viscosity: Concept and Measurement.* Elsevier Science.

Brookmire, L., P. Mallikarjunan, M. Jahncke and R. Grisso 2013. Optimum cooking conditions for shrimp and Atlantic salmon. *J. Food Sci.* 78(2): S303–S313.

Buransompob, A., J. Tang, R. Mao and B.G. Swanson. 2003. Rancidity of walnuts and almonds affected by short time heat treatments for insect control. *J. Food Proc. Pres.* 27(6): 445–464.

Chung, H., S. Wang and J. Tang. 2007. Influence of heat transfer in test tubes on measured thermal inactivation parameters for Escherichia coli. *J. Food Prot.* 70(4): 851–859.

Demirhan, E. and B. Özbek. 2009. Color change kinetics of microwave-dried basil. *Dry. Tech.* 27(1): 156–166.

Dolan, K.D. 2003. Estimation of kinetic parameters for non-isothermal food processes. *J. Food Sci.* 68(3): 728–741.

Engin, D. 2019. Effect of drying temperature on color and desorption characteristics of oyster mushroom. *Food Sci. Tech.* (Ahead).

FDA. 2000. Kinetics of microbial inactivation for alternative food-processing technologies. http://www.fda.gov/Food/FoodScienceResearch/SafePracticesforFoodProcesses/ucm100158.htm. Last accessed on May 2, 2014.

Fellows, P.J. 2009. *Food Processing Technology: Principles and Practice*, 3rd edn. Boca Raton: CRC Press, 895 pp.

Fu, B. and T.P. Labuza. 1993. Shelf-life prediction: Theory and application. *Food Contro.* 4(3): 125–133.

Fujikawa, H. and T. Itoh. 1998. Thermal inactivation analysis of mesophiles using the Arrhenius and z-value models. *J. Food Prot.* 61(7): 910–912.

Hadjal, T., C. Dhuique-Mayer, K. Madani, M. Dornier and N. Achir. 2013. Thermal degradation kinetics of xanthophylls from blood orange in model and real food systems. *Food Chem.* 138(4): 2442–2450.

Harbourne, N., J.C. Jacquier, D.J. Morgan and J.G. Lyng. 2008. Determination of the degradation kinetics of anthocyanins in a model juice system using isothermal and non-isothermal methods. *Food Chem.* 111(1): 204–208.

Hindra, F. and O.D. Baik. 2006. Kinetics of quality changes during food frying. *Crit. Rev. Food Sci. Nutri.* 46(3): 239–258.

Hiwilepo-van Hal, P., C. Bosschaart, C. Van Twisk, R. Verkerk and M. Dekker. 2012. Kinetics of thermal degradation of vitamin C in marula fruit (Sclerocarya birrea subsp. caffra) as compared to other selected tropical fruits. *LWT-Food Sci. Technol.* 49(2): 188–191.

Holdsworth, D. and R. Simpson. 2008. *Thermal Processing of Packaged Foods*. London: Blackie Academic and Professional, 427 pp.

Jaiswal, A.K. and S. Gupta and N. Abu-Ghannam. 2012. Kinetic evaluation of colour, texture, polyphenols and antioxidant capacity of Irish York cabbage after blanching treatment. *Food Chem.* 131(1): 63–72.

Jonsson, U., B.G. Snygg, B.G. HäNulv and T. Zachrisson. 1977. Testing two models for the temperature dependence of the heat inactivation rate of Bacillus stearothermophilus spores. *J. Food Sci.* 42(5): 1251–1252.

Kahyaoglu, T. and S. Kaya. 2006. Modeling of moisture, color and texture changes in sesame seeds during the conventional roasting. *J. Food Eng.* 75(2): 167–177.

Ko, W.C., W.C. Liu, Y.T. Tsang and C.W. Hsieh. 2007. Kinetics of winter mushrooms (Flammulina velutipes) microstructure and quality changes during thermal processing. *J. Food Eng.* 81(3): 587–598.

Kong, F.B., J.M. Tang, B. Rasco and C. Crapo. 2007. Kinetics of salmon quality changes during thermal processing. *J. Food Eng.* 83(4): 510–520.

Labuza, T.P. 1984. Application of chemical kinetics to deterioration of foods. *J. Chem. Educ.* 61(4): 348.

Lau, M.H., J. Tang and B.G. Swanson. 2000. Kinetics of textural and color changes in green asparagus during thermal treatments. *J. Food Eng.* 45(4): 231–236.

Liaotrakoon, W., N. Clercq, V. Hoed, D. Walle, B. Lewille and K. Dewettinck. 2013. Impact of thermal treatment on physicochemical, antioxidative and rheological properties of white-flesh and redflesh dragon fruit (Hylocereus spp.) purees. *Food Biopro. Tech.* 6(2): 416–430.

Liing, A.C. and D.B. Lund. 1978. Determining kinetic parameters for thermal inactivation of heat resistant and heat-labile isozymes from thermal destruction curves. *J. Food Sci.* 43(4): 1307–1310.

Ling, B., J. Tang, F. Kong, E.J. Mitcham and S. Wang. 2015. Kinetics of food quality changes during thermal processing: A review. *Food Biopro. Tech.* 8(2): 343–358.

Liu, Y., J. Tang, Z. Mao, J.-H. Mah, S. Jiao and S. Wang. 2011. Quality and mold control of enriched white bread by combined radio frequency and hot air treatment. *J. Food Eng.* 104(4): 492–498.

Lund, D.B. 1977. Design of thermal processes for minimizing nutrient retention. *Food Tech.* 31: 71–78.

Marra, F., L. Zhang and J.G. Lyng. 2009. Radio frequency treatment of foods: Review of recent advances. *J. Food Eng.* 91(4): 497–508.

Maskan, A., S. Kaya and M. Maskan. 2002. Effect of concentration and drying processes on color change of grape juice and leather (pestil). *J. Food Eng.* 54(1): 75–80.

Maskan, M. 2001. Kinetics of colour change of kiwifruits during hot air and microwave drying. *J. Food Eng.* 48(2): 169–175.

Nadia, D.M., Bonazzi, C., K. Nabil and M.B. Nourhène. 2011. Effect of air drying on color, texture and shrinkage of sardine (Sardina pilchardus) muscles. *J. Nutrition Food Sci.* 1: 113.

Nisha, P., R.S. Singhal and A.B. Pandit. 2006. Kinetic modelling of texture development in potato cubes (Solanum tuberosum L.), green gram whole (Vigna radiate L.) and red gram splits (Cajanus cajan L.). *J. Food Eng.* 76(4): 524–530.

Onwude, D.I., N. Hashim, R. Janius, N.M. Nawi and K. Ebdan. 2016. Color change kinetics and total carotenoid content of pumpkin as affected by drying tempearture. *Italian J. Food Sci.* 29(1).

Ovissipour, M., B. Rasco, J. Tang and S.S. Sablani. 2013. Kinetics of quality changes in whole blue mussel (Mytilus edulis) during pasteurization. *Food Res. Inter.* 53(1): 141–148.

Peleg, M., M.D. Normand and M.G. Corradini. 2012. The Arrhenius equation revisited. *Crit. Rev. Food Sci. Nutri.* 52(9): 830–851.

Peng, J., J. Tang, D.M. Barrett, S.S. Sablani and J.R. Powers. 2014. Kinetics of carrot texture degradation under pasteurization conditions. *J. Food Eng.* 122: 84–91.

Pereira, R., R. Martins and A. Vicente. 2008. Goat milk free fatty acid characterization during conventional and ohmic heating pasteurization. *J. Dairy Sci.* 91(8): 2925–2937.

Saguy, I., I.J. Kopelman and S. Mizrahi. 1978. Simulation of ascorbic acid stability during heat processing and concentration of grapefruit juice. *J. Food Proc. Eng.* 2(3): 213–225.

Saxena, A., T. Maity, P.S. Raju and A.S. Bawa. 2012. Degradation kinetics of colour and total carotenoids in jackfruit (Artocarpus heterophyllus) bulb slices during hot air drying. *Food Biopro. Tech.* 5(2): 672–679.

Simpson, R. 2010. *Engineering Aspects of Thermal Food Processing*. Boca Raton: CRC Press, 522 pp.

Song, F., R.D. Cannon and M. Bochmann. 2003. Zirconocene-catalyzed propene polymerization: A quenched-flow kinetic study. *J. Amer. Chem. Soc.* 125(25): 7641–7653.

Steinfeld, J.I., J.S. Francisco and W.L. Hase. 1998. *Chemical Kinetics and Dynamics*. Upper Saddle River: Prentice Hall, 560 pp.

Stoforos, N.G. 1995. Thermal process design. *Food Cont.* 6(2): 81–94.

Tang, J., H. Feng and M. Lau. 2002. Microwave heating in food processing. In: X. Young and J. Tang (eds.), *Advances in Bioprocessing Engineering*. New Jersey: World Scientific Publisher, pp. 1–43.

Theodore, L., I.S. Saguy and T. Petros. 1997. Kinetics of food deterioration and shelf-life prediction. In *Handbook of Food Engineering Practice*. New York: CRC Press.

Toledo, R.T., R.K. Singh and F. Kong. 2007. *Fundamentals of Food Process Engineering* (Vol. 297). New York: Springer.

Van Boekel, M.A.J.S. 1996. Statistical aspects of kinetic modeling for food science problems. *J. Food Sci.* 61(3): 477–486.

Van Boekel, M.A.J.S. 2002. On the use of the Weibull model to describe thermal inactivation of microbial vegetative cells. *Inter. J. Food Micro.* 74(1–2): 139–159.

Vieira, M.C., A.A. Teixeira and C.L.M. Silva. 2000. Mathematical modeling of the thermal degradation kinetics of vitamin C in cupuaçu (Theobroma grandiflorum) nectar. *J. Food Eng.* 43(1): 1–7.

Vikram, V.B., M.N. Ramesh and S.G. Prapulla. 2005. Thermal degradation kinetics of nutrients in orange juice heated by electromagnetic and conventional methods. *J. Food Eng.* 69(1): 31–40.

Wang, J., K. Luechapattanaporn, Y. Wang and J. Tang. 2012. Radiofrequency heating of heterogeneous food—meat lasagna. *J. Food Eng.* 108(1): 183–193.

Wedzicha, B.L., S.J. Goddard and A. Zeb. 1993. Approach to the design of model systems for food additive-food component interactions. *Food Chem.* 47(2): 129–132.

Wen, T.N., K.N. Prasad, B. Yang and A. Ismail. 2010. Bioactive substance contents and antioxidant capacity of raw and blanched vegetables. *Ino. Food Sci. Emer. Tech.* 11(3): 464–469.

Yang, J., G. Bingol, Z. Pan, M.T. Brandl, T.H. McHugh and H. Wang. 2010. Infrared heating for dry-roasting and pasteurization of almonds. *J. Food Eng.* 101(3): 273–280.

Yoon, Y., W.J. Cho, J.G. Park, J.N. Park, B.S. Song, J.H. Kim, M.W. Byun, C.J. Kim, A.K. Sharma and J.W. Lee. 2009. Effect of gamma irradiation on shelf-life extension and sensory characteristics of Dak-galbi (marinated diced chicken) during accelerated storage. *Korean J. Food Sci. Ani. Res.* 29(5): 573–578.

Zabbia, A., E.M. Buys and H.L. De Kock. 2011. Undesirable sulphur and carbonyl flavor compounds in UHT milk: A review. *Cri. Rev. Food Sci. Nutri.* 52(1): 21–30.

Zanoni, B., E. Pagliarini, G. Giovanelli and V. Lavelli. 2003. Modelling the effects of thermal sterilization on the quality of tomato puree. *J. Food Eng.* 56(2–3): 203–206.

2 Advances in Drying Technology for Food Processing

Ipsita Das
Centre for Technology Alternatives for Rural Areas, Indian Institute of Technology Bombay, Powai, Mumbai, India

S.K. Das
Department of Agriculture and Food Engineering, Indian Institute of Technology Kharagpur, West Bengal, India

CONTENTS

2.1 FUNDAMENTALS OF DRYING

Drying or dehydration of food grains under the sun is the oldest method of food preservation, adopted since the beginning of human civilization. Dehydration is the removal of moisture from foods, causing a reduction in the activity of microorganisms and the chemical potential of water molecules (water activity) (Rizvi, 2005). Water exists in free and bound forms in foods. Free water exists as a liquid or unbound water (water activity is 1), and bound water exists in association with the sorbent or solute. Bound water differs thermodynamically from liquid water (water activity less than 1) and exhibits a decreased diffusion coefficient when the moisture content of the food is reduced. As a result, water reaches the surface more slowly, which impedes drying. Water transport mechanisms, process kinetics, and energy requirements in the dehydration process are governed by a number of factors, including the energy of the sorption sites, steric configuration, solute–sorbent interactions, pH, temperature, etc. These factors are decisive for various quality attributes of dehydrated foods.

In practical terms, drying of foods is a process in which both unbound and bound moisture is removed to the surrounding environment until a product with the desired, or low, moisture level is obtained. With a prolonged drying process, the food will attain equilibrium moisture content. A simultaneous process of heat and mass transfer occurs in drying: heat supplied from the surroundings to the surface of the object being dried is utilized for evaporation of water, and the internal liquid water in the food is transported to the surface by molecular diffusion. Water removal as vapor from the surface to the environment is a function of the physical condition of the material (area exposed to drying) and the external conditions (temperature, humidity, airflow, and pressure). Moisture movement inside the material basically depends on two factors: the physical structure of the material and the moisture content (unbound and bound). The temperature of the material also influences the diffusion of moisture. Any of these parameters may limit the drying rate and the final moisture content of the product.

In conventional drying systems (hot air in contact with the solid surface), heat transfer occurs from the surface of the wet solid to the interior. However, during microwave or radiofrequency drying, the energy supplied is utilized for heat generation within the solid body (volumetric heating). The urgent need for energy management and the demand for better-quality dried foods have led, over the last 20 years, to extensive research on novel alternative dehydration techniques for materials in general and foods in particular, on both the theoretical and applied fronts.

2.1.1 Development of Drying Technologies

Our ancestors learned about drying of foods by trial and error. Since ancient times, cereal grains, fruits, vegetables, fish, and meat have been dried using sunlight and smoke. Around 5000 years ago, Persians and Chinese people learned to dry fruits and vegetables under the sun (Ratti, 2009). About eight centuries ago, Incas in the Andean highlands (pre-Columbian America) developed a natural way of freeze-drying potato (called *chuño*) that required almost no technology (Micu, 2016). Interestingly, this technique can still be found in such applications as foods for astronauts who require extended shelf life, diet supplements, and easy transportation. Drying of raisins, figs, and apricots under the sun is thought to have been developed by aboriginal inhabitants of the Mediterranean Basin and Near East (Ratti, 2009).

While natural drying of food grains using sun and wind developed long ago, the first artificial drying of foods dates from the eighteenth century, as reported by van Arsdel and Copley (1963). World Wars I and II accelerated the growth of drying industries, with the need for large-scale production for the military. A series of techniques for drying solid (fruits and vegetables, grains) and liquid foods (milk, eggs, etc.) were developed for the commercial production of dehydrated products. These included drum drying, spray drying, vacuum drying, and freeze-drying. Production of dehydrated foods evolved from processes developed for wartime uses to a vigorously growing branch of the entire food processing industry (Barbosa Canovas and Vegas-Mercanto, 1996). Numerous dehydration technologies, including microwave drying, heat-pump drying, and superheated steam drying, emerged during the last part of the twentieth century. According to Mujumdar (2015), around 400 types of dryer have been reported, over 100 of which are commonly available.

Currently, growing consumer demand for quality food products and eco-friendly sustainable processes necessitate improvement of the performance of existing drying technologies and the development of drying systems with new concepts (Sabarez, 2016). Table 2.1 shows the stages of development of food dehydration systems.

2.1.2 Criteria for Selection and Classification of Dryers

The diverse nature of foodstuffs necessitates the use of different types of dryers and drying techniques. A number of authors (Kudra and Mujumdar, 2002; Mujumdar and Menon, 1995; Vega-Mercado et al., 2001) have classified drying systems and industrial dryers based on their mode of operation (batch, semi-continuous, continuous), drying medium (air, flue gas, steam, solvent), operating pressure (sub-atmospheric, normal or above atmospheric pressure), operating temperature (above boiling point, below boiling point, sub-freezing), mode of heat transfer (conduction, convection, radiation, combination), nature of the feed material (solid, liquid, paste), state of the materials being dried (granular, shaped, emulsion, fine spray, layer), movement of the materials being dried (stationary, moving, agitated, dispersed, fluidized), relative flow of drying medium and the material to be dried (co-current, counter-current, mixed),

TABLE 2.1
Development of Food Dehydration Systems

Early civilization	• Process of drying was known for extending storage life of foods. • Sun for energy, wind for water transport, and ample open space to spread were basic criteria for drying of foodstuffs. • Weather dependency and slow drying were tackled by placing foods on hot stone slabs (heated from below by burning wood, leaves, and dung) till they attained low moisture level.
Mid-civilization	• Labor-intensive process, less open area, and slow drying triggered the invention of simple mechanical dryers. • Dryers using chimneys and flue gas (burning wood, leaves) for direct heating were installed, subsequently modified with indirect heating (kiln dryers). • Better mass transfer was achieved with hot-air blowing using manually operated bellows.
Late 19th century	• Design of simple dryers for hot-air drying of foods was initiated. Mechanical dryers with hot air and blowers came into use with the invention of the electric motor and blowers.
Early 20th century	• Role of water in the deterioration of foods and extension of shelf life by reducing their water content became well understood from a thermodynamic standpoint **(water activity)**. • Moisture sorption isotherms and MSI models were used as tools for dryer design and prediction of the storability of foods. • Rapid development of various dryers for dehydration of almost all kinds of solid and liquid foods using conduction, convection and thermal radiation heat-transfer systems occurred, and commercial production of dehydrated foods began in response to consumer demand.
Late 20th century	• Various issues related to the development of energy-efficient, compact and versatile dryers were addressed in international forums, leading to replacement or supplementing of existing drying methodologies and dryers, or hybridization.

length of time in the drying process (short, medium, long), and stages of drying (single, multistage, re-circulatory) (Kudra and Mujumdar, 2002). Other selection criteria for the classification of dryers (Mujumdar and Menon, 1995) are based on product sensitivity to heat and oxidation, final product quality, dryer throughput, cost of operation (including nature of fuel to be used), environmental pollution and other hazards, convenience of installation and operation, space constraints, and the requirement for precise control of drying parameters. Drying kinetics and thermodynamics of the desorption process are also guiding factors in the selection of drying systems.

2.1.3 Issues with Current Drying Techniques for Foods

Specific energy consumption is a measure of dryer efficiency, expressed as the ratio of total heat input to the dryer to the mass of water evaporated. This aspect is very important for the commercial adaptability of the dryer and closely related to the drying rate. A number of other factors are associated with current conventional drying systems:

- Space requirements and ease of maintenance
- Quality of the dried products

- Safety aspects in operation
- Optimum performance with significant changes in throughput
- Control of drying parameters with proper instrumentation
- Environmental impact due to combustion of fossil fuels
- Versatility in operation

Many existing conventional drying technologies developed empirically. The current energy utilization pattern has now become binding for food-processing systems utilizing these dryers and drying systems. A continuous production system requires automated online quality control with appropriate instrumentation. In addition, environmental pollution and finished product quality regulations are becoming more stringent and binding.

2.1.4 Need for Advanced Drying Technology for Foods

One of the major concerns with conventional food drying systems is that their performance level has reached the asymptotic limit. Dynamic upgrading of many current systems is in keeping with legislative and competitive constraints. The main considerations in development of alternative or new food drying technologies are cost-effective performance increase (as well as higher asymptotic performance limit) and increased productivity.

Reducing the unit size helps reduce the initial capital investment, and space, ducting, and ancillary equipment requirements. Dryer volume is closely related to the prevailing drying rate in the system. Intensified heat and mass transfer result in high volumetric evaporation rates and, consequently, a high drying rate. Ultimately, dryer performance can be significantly increased by intensifying the drying rate. However, the effect of such measures on product quality should be investigated.

The rate of drying with a convective dryer does not solely depend on the material being dried. Heat and mass transfer at the external surface of the material may be enhanced by changing the operating parameters, including air velocity, temperature, and turbulence of the free-stream air, intense mixing or turning, oscillation of flow, and decreased relative humidity of the air. A significant drop in the drying rate during the falling-rate stage can be attributed to slow moisture transport from the interior of the material to its surface (resistances in heat and mass transfer). The following approaches can be used to achieve faster heat and moisture transport:

- Use of infra-red to enhance penetration of heat into the product, which results in faster moisture transport from the interior to the surface of the material. Application of an ultrasonic field generates high-frequency pressure pulsation (successive high-pressure and low-pressure fields), resulting in cavitation, which improves liquid transport from inside the material due to rapid vaporization.
- Application of microwave or radio-frequency energy causes volumetric heating of water (polar material) present inside the material, dramatically reducing internal resistance to heat transfer. The high inside temperature results in high diffusivity of moisture in the solid matrix. Along with the internal pressure gradient, it speeds up moisture transport from inside to the surface.

- Low-pressure field (vacuum) decreases the boiling temperature of the water inside the food, enhancing the transport of moisture by rapid evaporation.
- Mechanical dispersers or mixers increase the exposed area of the material being dried.
- An impinging flow configuration (two-phase solid–gas flow) increases the evaporation of surface moisture by a factor of 2–3.
- Mixing hot inert particles with wet particles results in good particle-to-particle heat and mass transfer leading to very high drying rates.
- Using adsorbent particles in the drying increases heat and mass transfer with the heat of sorption and hygroscopicity.

2.2 ADVANCED DRYING TECHNIQUES

Several advanced drying techniques are described in the following sections.

2.2.1 Contact–Sorption Drying

Contact–sorption drying is an amalgamation of two drying techniques: classical contact drying, in which heat conducted from a heated solid surface is utilized for evaporation of moisture; and sorption drying (also termed desiccant drying or adsorption drying), in which the concentration gradient between the adsorbent material and the material being dried is the factor for moisture transfer.

In this drying system, solid desiccant material is mixed with the wet material undergoing drying. Mixing continues till the desired level of mass transfer is reached and is followed by a suitable separation process. One of the important technical parameters is regeneration of the solid desiccant material (sorbent), which is then recycled. The success of this process depends on the ease of regeneration. The sorbent may be classified as passive (inert) or active. In the passive category, the sorbent material is used to remove moisture from the wet material, followed by separation from the dried material. In the active category, the sorbent material becomes an integral part of the final dried material. Typical examples of inert sorbent materials are silica gel, molecular sieves, activated carbon, zeolite, chabazite, and bentonite. Active sorbent materials may be of three types: (i) extraneous material; (ii) material belonging to the same product as the one being dried; or (iii) a carrier or filler. Examples of extraneous materials are starch, bran, straw, cellulose, and sugar beet pulp. If the active sorbent used belongs to category (ii), the sorbent and the wet feed are mixed before drying. A good example of this category is the simultaneous drying and granulation of biomaterials in a fluidized-bed dryer. Sorbent materials chosen as carrier or filler are usually used in solution or suspension, and are either spread over the surface of the drying particles or absorbed in the porous structure of the solid material (Kudra and Mujumdar, 2002).

2.2.1.1 Contact–Sorption Drying Mechanism

Moisture can be removed from wet food materials if a highly hygroscopic material (sorbent) is mixed with the wet feed. Moisture content of the wet feed is gradually reduced in the particle-to-particle contact mass transfer process. Heat and mass transfer occur simultaneously on contact between heated sorbent particles and the wet solid, increasing the drying rate significantly (Kudra and Mujumdar, 1995).

2.2.1.2 Moisture Transfer by Different Mechanisms

a. **Suction stage**

   The capillary flow mechanism causes adsorption of liquid moisture (from the wet feed) by the sorbent material. The moisture transfer rate depends significantly on the contact area. Moisture transfer at the interface stops when the suction potential at both surfaces becomes equal (equilibrium condition).

b. **Diffusion stage**

   The developing concentration gradient causes moisture diffusion in both components: for the sorbent moisture diffuses from the surface to the entire volume, and for the material being dried, moisture diffuses from the core to surface.

c. **Diffusion with inter-particle collisions**

   Sorbent–sorbent and material–material contacts occur in fluidized beds, vibrated fluidized beds, and spouted beds. This phenomenon leads to additional moisture transfer at the microscopic level.

Relative sorption capacity of sorbents describes their specific sorption capacity, under comparable conditions, compared to that of the material under reference (Kudra and Mujumdar, 2002). Thus, the sorption characteristics of the adsorbent material determine the efficiency of contact–sorption drying, selection of the drying method, and design of the contact–sorption dryer. The specific sorption capacity at a given relative air humidity is expressed as the ratio of the mass of water adsorbed to the mass of a sorbent taken. Another way of expressing relative sorption capacity is in terms of the moisture content. A filter paper is taken as reference material because it possesses constant sorption capacity over a wide or the whole range of humidity. Materials with K less than 1.0 are considered low-capacity sorbents. Materials having K-value greater than 3.0 are considered high-capacity sorbents.

2.2.1.3 Categorization of Sorbents Based on K-Values

Depending on the specific range of relative humidity values of the system, sorbents are categorized as low-, moderate-, and high-capacity sorbents. Examples of high-capacity sorbents at high ambient humidity range (80–90%) are wheat bran, silica gel, and ground rapeseed. All these materials are suitable for spray drying or fluidized-bed drying systems. Zeolites and activated carbon have high K-values at a low vapor pressure environment (low RH). Thus, they are best in freeze- and vacuum-drying processes where there is low relative humidity. Typical examples of contact–sorption drying systems are continuous drying of particulate materials over inert sorbent, and a contact–sorption fluid-bed dryer for pasty materials.

The continuous process consists of three long coaxial, conical-shaped drums with typical dimensions of 1.2 m length and 0.8 m diameter, and an angular speed of

100–160 rpm. The (gas-fired) heated innermost drum directly heats the inert particles (200–300°C) while they traverse the heating chamber. The chamber wall consists of scope-like staggered flights, which facilitates the showering of particles and better heating. Finally, the hot particles are discharged at the other end. Wet particulate material (feed) enters the dryer (an annular space between the heated drum and the outer drum), mixes with these heated particles and moves together with the heated particles with the help of an auger located between the outer conical drum and the heating chamber. It takes a short contact time (15–20 s) with these hot particles for the moisture to evaporate from the wet material. The tail end of the outermost conical chamber is perforated, enabling separation of inert particles, and finally, the collected particles are recycled. The dry materials (bigger in size) are discharged at the far end.

Pasty and sticky materials are dried with fluidized fine active sorbent particles using a two-stage dryer. The upper part is a fluidized dryer into which granular wet material falls and is coated with the hot sorbent particles. The surface of the granules is desiccated rapidly due to the high heat and mass transfer rate, and a non-sticky (non-aggregating) and porous surface is formed where moisture is easily diffused. In a second stage, the granules are dried in a second fluidized dryer to obtain the desired final moisture content.

2.2.2 Dielectric Drying

Dielectric drying technologies using microwave (MW) and radio frequency (RF) as energy source have emerged in recent decades. A frequency between 10 and 300 MHz is used in RF heating. For MW heating, the frequency range is wide – between 300 MHz and 300 GHz. However, the specific frequency ranges chosen for RF are 13.56, 27.12, and 40.68 MHz, respectively for domestic, industrial, scientific and medical uses. The frequency values in MW heating are fixed as 915 (896 in Europe) and 2450 MHz (Mujumdar, 2000).

Unlike conventional thermal processing using conduction, convection, and radiation for heat transfer, dielectric heating primarily involves two mechanisms: (1) dipolar rotation, in which the polar molecules (such as water in food) try to orient themselves with the frequency of the changing electric field; and (2) ionic conduction, in which the hydrated ions move to and fro inside the sample with the changes in the electric field. In the RF range, ionic conduction is more significant; both ionic conduction and dipole rotation are important in the RF range (Strumillo and Kudra, 1986). Foodstuffs undergo rapid heating as a result of volumetric heat generation with the application of RF or MW energy. The primary factor for MW- and RF drying processes is the dielectric properties (ε) of the foods. Foodstuffs exhibit very complicated dielectric spectra, which are a function of frequency, temperature, and time. The spectra ultimately depend on the dipolar frequencies of the polar molecules within the foodstuff (Daschner et al., 2000).

Dielectric properties are expressed in terms of dielectric constant (ε′) and dielectric loss factor (ε″) as:

$$\varepsilon = \varepsilon' - j\varepsilon'' = |\varepsilon| e^{-j\delta}$$

The dielectric constant of the material concerned is associated with its ability to store electrical energy. The dielectric loss factor indicates the dissipation of stored electrical energy. The term δ in this equation is known as dielectric loss angle [$\delta = \tan - 1(\varepsilon''/\varepsilon')$]. Physically, it indicates the ability of a compound to convert the absorbed microwave radiation into thermal energy.

2.2.2.1 Microwave Heating

Microwave (MW) heating is a relatively mature technology in food drying which gives accelerated drying rates significantly compared to conventional solar or hot-air drying methods. MW drying enables precise control over the process and acceptable quality for the final dried products (Jiang et al., 2010). However, some limitations, such as non-uniform temperature distribution, limited penetration depth, and textural damage have not yet been fully resolved to enable this drying method to be used on its own. Combining MW drying with other techniques, such as hot-air drying, vacuum drying, freeze-drying or spouted-bed drying, has been suggested. Many studies have investigated MW drying, as well as hybrid drying for various food and agricultural products including fruits (mango, blueberry, banana, apple), vegetables (carrot, green soybean, potato, garlic) and aquatic products (Jiang et al., 2017). Wang et al. (2014) worked with MW-assisted drying and reported a reduction in both drying time (25–90%) and energy consumption (32–71%) compared to conventional drying techniques. However, the practical industrial implementation of MW drying presents some challenges in scaling up, especially as regards the penetration of energy and uniformity in heating.

2.2.2.2 Radio Frequency (RF) Heating

Although RF drying is used around the world commercially in the wood, textile, and paper industries, it is not used for drying of foods on an industrial scale. Its main uses in foods are drying of pasta and post-bake drying of cookies and crackers. RF drying has the advantage of higher penetration depth than MW drying. The longer wavelength is also more suitable for bulk as well as uniform heating of the materials (Vega-Mercado et al., 2001). MW heating of milk or yogurt products using 2.45 kHz showed a penetration depth of about 1 cm, but when RF of 27.12 MHz was used, it was 20 cm (Felke et al., 2009). RF drying systems are quite costly for drying of large-volume materials, such as cereal grains. Nevertheless, for high-value products, the quality of the final product can be improved, especially in situations where conventional drying techniques are unsuitable. Despite significant development work, commercialization of RF/MW drying of foods is quite slow to progress.

RF drying has a number of limitations. It is highly capital intensive: the conversion efficiency of electrical energy for generating a dielectric field in the RF heating process is only around 50% (Mujumdar, 2000). Non-uniformity of heating makes the drying process difficult, particularly for irregular-shaped heterogeneous foods. Engineering problems include the short life of the energy conversion devices. Finally, exhaustive data on dielectric properties of different foods are required for the successful operation of the system.

2.2.3 Drying with Inert Solids

This technique of drying solutions, slurry, or paste (pigments, chemicals, and biological materials) to make dry powder was developed in the former Soviet Union in 1950. This technology is attracting renewed interest as an alternative to spray, drum, and film rotary drying for the conversion of dispersed solids in the liquid to dry powder (Kudra and Mujumdar, 1995; Strumillo et al., 1983). Typically, fluidized-bed drying of slurry using inert particles showed 15–17 times higher volumetric evaporation than standard spray drying under identical thermal conditions (Kudra and Mujumdar, 1995). Following extensive research, a pilot commercial-scale fluid-bed dryer and swirling-bed dryer have been developed using inert particles (Grbavcic et al., 1998; Kutsakova, 2007).

2.2.3.1 Drying Mechanism

Liquid feed is commonly dispersed on the surface of inert solid particles (a carrier) using mainly fluidized-bed, swirling, impinging, and pneumatic dryers. Fluidized-bed drying systems include simple fluidized-bed, spouted-bed, spout-fluidized-bed, jet-spouted-bed, vibro-fluidized-bed, and cyclone dryers. The inert particles in the drying bed are fluidized as a result of the hydrodynamic impact of the air stream (Kudra and Mujumdar, 2002). Fluidization by impact can also be achieved through the combined action of airflow and mechanical vibration. An oscillating external magnetic field applied to ferromagnetic inert particles can also result in fluidization. Drying is accomplished by the transfer of heat from the air by convection, and from the stored heat in the particles by conduction. The dried mass on the particles is peeled off due to particle–particle collision. The dry material is transformed into fine powder or small flakes due to intense attrition and discharged with the exhaust air. Finally, the dried product is collected.

The kinetics of drying with inert particles comprise four stages: (i) heating; (ii) coating with dispersed liquid; (iii) drying of the coating material; and (iv) breaking and peeling off the final dried product. Coating with liquid occurs simultaneously with peeling of the dried material; the latter operation should always be greater than the former, otherwise, the drying system will fail. The efficiency of the drying process is dependent on the nature of the particles and the physical properties of the coating material (Kudra and Mujumdar, 2009).

2.2.3.2 Vibro-Fluidized-Bed Dryer Using Inert Particles

Drying of pasty materials with a vibro-fluidized dryer using inert particles (cubes of polytetrafluoroethylene) is performed in a vertical cylindrical chamber, into which two horizontal pneumatic nozzles at opposite sides spray the material. The particles vibrate mechanically on a perforated support in addition to fluidization by the flow of hot air from the bottom (Rysin and Ginzburg, 1992). The spray material makes a coating (film) on the surface of the inert particles. Rapid drying of this resulting film occurs with the combined effects of heat conduction from the inert particles, and heat and mass convection with the air flowing around the particles. The dry film disintegrates into powder due to intense high-frequency impacts between the particles, and the powder carried in the airstream is separated by a cyclone separator. This process proceeds in dynamic equilibrium after a brief period of unsteady-state operations. Materials having low adhesiveness towards the inert particle, such as egg-white, egg-yolk, animal blood, casein, yeasts, and soya protein, are suitable for this kind of drying. Materials unsuitable for this drying technique are whole and skimmed milk, coffee, tea, protein-hydro lysates, fruit pulps, etc. These materials possess high adhesiveness towards the inert particles.

2.2.4 Pneumatic Flash Drying

In pneumatic flash drying (or convective drying), there is direct contact between the solid particles and the drying medium (heated gas stream), which also acts as a transport medium for the solid particles. The particles (usually in the range of 10–500 μm) attain the desired moisture content within a short time in the drying section. Higher drying rates and higher drying capacity are achieved with the large surface area in these units and the high convective heat and mass transfer coefficients (Levy and Borde, 2015).

Pneumatic flash drying consists of three basic units: gas heating and ducting unit, material feeding unit, and product collection unit along with a recirculation line for recycling of materials for proper drying. The wet feed enters the feeder which has a special mixing device. It is then carried through the tubes in air flowing faster than the free-fall velocity. Finally, it is separated in a cyclone (Dincer and Zamfirescu, 2016). This drying system removes surface moisture more quickly than the moisture can be transported from the interior part. Moisture transport is controlled by diffusion. Simple construction (fewer moving parts), low capital investment, compact design (small installation area), easily controllable features, simultaneous drying and carrying of materials, and almost trouble-free operation (low maintenance) make this pneumatic flash dryer very useful industrial equipment for the drying of granular, pasty, and powdery products. This system also makes use of the low air temperatures that favor the drying of thermo-labile materials. The disadvantages with this drying system include low drying efficiency with sticky or lumpy materials, risk of fire and explosion, and the requirement for high efficiency of the product collection (separating) unit.

2.2.5 Foam-Mat Drying

Foam-mat drying is used to dry liquid foods in which the liquid is pre-concentrated (optional, depending on initial viscosity and surface tension) and then transformed into stable foam. This foam is layered onto a surface and subjected to continuous drying under normal atmospheric pressure using dehumidified air at relatively low temperatures. The drying process is relatively faster with an enlarged surface area, and the dried foam is finally transformed into a thin porous sheet or mat with a honeycomb structure. The free-flowing powder is easily obtained after the disintegration of the sheet or mat, using a suitable size-reduction system. This drying process at lower temperatures is suitable for making juices into powder. Nutritional quality retained in the final product is good, and it can be easily reconstituted. Air drying can be replaced by conduction-type refractive window drying (see below). According to some authors (Saravacos and Kostarapoulos, 2002; Kadam et al., 2010; Febrianto et al., 2012), foam-mat drying is considerably cheaper than vacuum, freeze and spray drying. However, large-scale spray dryers are more efficient for commercial applications, are the most suitable method of drying food products that are heat sensitive, viscous, sticky, and high in sugar content (Rajkumar et al., 2007). Spray or drum drying of these kinds of products presents some difficulties.

2.2.5.1 Foam Formation and Stability

Foam is a stabilized three-phase gas–liquid–gas system. Foam texture is dependent on the number, size, and distribution of bubbles per unit volume. Surfactants such as proteins (egg-white, soyproteins) and other synthetic compounds (glycerol-monostearate, carboxymethyl-cellulose) give kinetic stability during the formation of foam due to modification of the inter-particle forces between the foam bubbles (Indrawati et al., 2008), as well as influencing its texture. Concentration of foaming agent, solid content in the mass, time of agitation, and temperature also affect the foaming properties of these proteins. The surfactants bind at the interface and interact at the thinner regions of the bubble walls (called lamella), facilitated by electrostatic or hydrophobic forces, hydrogen bonds, or covalent linkages. The formation of foam depends on properties of the liquid, foaming method, and conditions maintained during foaming. As well as the type of surfactant, various processing parameters, such as rate of mixing, temperature, and geometry of the mixing impeller, affect the volume of air incorporation in the foam, bubble size, and foam stability (Indrawati et al., 2008).

2.2.5.2 Foam Generation Methods

Whipping, shaking (rarely used), and bubbling are the static methods generally used to make foam with liquid foods (Hardy and Jideani, 2017).

Whipping

Devices such as blenders (manual and automatic), homogenizers, and vortex mixers are used to form foam. Increasing the intensity of beating or whipping increases the volume of air incorporated into the foam. The final size of bubbles depends on various factors, including rheological properties of the liquid, agitator speed, and geometry of the vessel (Arzhavitina and Steckel, 2010).

Shaking

Vigorous agitation of the liquid and air in a closed vessel results in formation of a low volume of foam. Frequency and amplitude of shaking of the mixture, volume and shape of the vessel, and volume and viscosity of the liquid determine the rate of formation of bubbles in the foam (Arzhavitina and Steckel, 2010).

Bubbling

Passing a swarm of air bubbles through a fine sparger immersed in the liquid produces foam. The size of the bubbles is highly dependent on the viscosity of the liquid, diameter of the opening, amount of foaming agent, and volume of liquid in the vessel (Arzhavitina and Steckel, 2010).

2.2.5.3 Foam-Mat Drying Technique

An important criterion for foam-mat drying is that the foam should remain stable during drying. Although the large volume of gas present in the foam mass does not favor heat transfer, the foam-mat technique achieves high drying rates, due to moisture movement with the capillarity (formed in the liquid films separating foam bubbles), large interfacial area and open structure of the foam (Rajkumar et al., 2007). During drying, bubbles continuously bursting in the foam change heat and mass transfer coefficients, resulting in an uneven constant-rate drying period. Foam-mat dried products possess good rehydration properties and retain their volatiles well (Kudra and Ratti, 2006).

2.2.5.4 Drying Process

Foam-mat drying is used for small-scale drying of sensitive food products such as concentrated fruit juice, fruit purees, and food slurries. The following sequence of steps is followed in foam-mat drying:

a) Pre-concentrating liquid foods (fruit and vegetable extracts) or adding viscosity enhancer.
b) Adding foaming agents or foam stabilizers.
c) Spreading the foam on trays or conveyor belts (about 1.5 mm deep).
d) Drying the foam using hot-air circulation maintained at 65–85°C with moderate air velocity.
e) Scraping dried material from the trays.
f) Grinding the porous dried mat to make free-flowing powders.

2.2.6 Heat-Pump Drying

Heat pumps, an efficient method of energy recovery, have been used since the middle of the twentieth century. Their application in various sectors, including dairy, fruits, grain and vegetable drying, have been reported. Heat-pump dryers (HPDs) have been developed by simple modifications of existing refrigeration systems which work on the principle of the thermodynamic cycle (i.e., extracting energy from a low-temperature heat source and delivering it to a high-temperature sink) and do not require much capital investment. Heat-pump dryers are a part of the vapor-compression, refrigeration-cycle system. The basic difference between the HPD system and the refrigeration system is an additional dryer, i.e., compressor, evaporator, condenser, and dryer. Air passing over the evaporator is cooled and dehumidified. Its relative humidity is then decreased by absorbing heat from the condenser. Thus, the evaporator acts as a dehumidifier, saturating the flowing air at low temperature. Heat from the condenser increases the difference between the dry-bulb and wet-bulb temperatures (called wet-bulb depression), and hence, the air has a lower relative humidity. The temperature and relative humidity of the output air can be well controlled. Controlled heat-pump drying is used to dry delicate or sensitive foods at regulated humidity and low temperatures, important for quality retention in dried products (Hawlader and Jahangeer, 2012).

The thermal characteristics of an HPD are its moisture extraction and specific moisture extraction rates. The former is defined as mass of moisture removed per unit time (kg h^{-1}); the latter is defined as mass of moisture removed per unit energy supplied (kg J^{-1}). Another characteristic of the HPD is coefficient of performance, which is defined as the amount of heat rejected by the condenser per unit of work done by the compressor, indicating the energy efficiency of the heat pump (Fayose and Huan, 2016). The major advantages of heat-pump dryers over conventional drying systems are improved energy efficiency, high moisture extraction rate, well controlled drying conditions, improved final product quality, ability to control drying parameters such as temperature and humidity, thereby creating a wide range of drying conditions, reduced electrical energy consumption, reduced operating cost and environment-friendliness. The temperature and humidity ranges generated with HPDs are 20–100°C and 15–80% respectively (Chua et al., 2002). Typically, drying efficiencies are reported to be ~95%, <70%, and 35–40% for HPD, vacuum drying, and hot-air drying respectively.

According to Perera and Rahman (1997), specific moisture extraction rate values ranged from 1.0–4.0, 0.72–1.2, and 0.12–1.28 kg·kW^{-1}·h^{-1} for HPD, vacuum drying, and hot-air drying, respectively. More recently, hybrid heat-pump systems (microwave, infrared, RF) have been developed to improve process efficiency by increasing drying rate while reducing thermal load. A 30–40% reduction in total energy consumption and adequate reduction in drying time were observed for the drying of saffron, apple, and mango slices (Mortezapour et al., 2012). Drying of herbs, mint leaves, ginger, and olive leaves using hybrid HPDs have also been reported (Moses et al., 2014). Heat-pump drying has been suggested as the best method for producing intermediate-moisture foods in fish processing (Shi et al., 2008).

Despite its promising potential, there remains significant scope for performance improvement of heat-pump drying systems using multiple heat pumps or heat pumps in cascade, heat pipes, absorption refrigeration, chemical heat pump, and so on (Baker, 1997), although the process is costlier. The HPD system should be used in the initial rather than the later stages of drying. Using dehumidified air in the later stages of drying has no significant effect, thus both capital and operating costs are reduced.

2.2.7 Impinging Stream Drying

Impinging stream drying (ISD) is a new kind of flash dryer. The dispersed wet particles or liquid droplets in a high-speed gas stream impinge on another high-velocity gas stream in the opposite direction, and drying occurs in the collision zone. Particles in the original stream are accelerated, penetrate the impingement plane, and finally decelerate to zero velocity after a certain distance of penetration. Thereafter, particles are again accelerated in the opposite stream, penetrate the impingement plane, and stop. The damped oscillation of the particles within this cycle of acceleration and deceleration causes a drop in the particle velocity in the impingement plane to the terminal velocity. Finally, the outgoing stream removes the particles. Larger particles remain longer in the impingement zone than in the gas stream. Inter-particle and non-elastic collisions, particularly where particle concentration in the feed is high, cause significant loss of particle momentum. The reduced particle residence time considerably reduces the beneficial effects of impingement. This problem can be alleviated by intermittent feeding of particles into the carrying stream (Kudra and Mujumdar, 2002).

As an alternative, impinging streams with a mobile impingement zone can be used so that the impingement plane moves between two distinct locations. The gas flows are alternated: from right to left, then left to right. The particles can be fed either into the accelerating flow duct or the central (reverse flow) duct linking the two impingement chambers (Kudra and Mujumdar, 2002).

Several variants of the impinging system can be used for drying of wet materials, depending on flow direction and flow geometry. There are eight types of flow: (1) coaxial, (2) mutually perpendicular, (3) curvilinear counter-current, (4) semi-circular or chain configuration, (5) counter-rotating counter-current, (6) co-rotating counter-current, (7) swirling, and (8) curvilinear. The geometries of the impingement plane/zone may be tubular, annular, or planar with radial/circumferential flow.

In general, heat transfer coefficients for ISDs have been reported to be much higher than for classical dryers with similar hydrodynamic regimes. Under certain operating conditions, for example, the heat transfer coefficient values are 850 $Wm^{-2}K^{-1}$ and 300–520 $Wm^{-2}K^{-1}$ in a coaxial ISD and pneumatic dryer, respectively (Strumillo and Kudra, 1986). Heat transfer coefficients 1.1–1.8 times higher have been reported for a dryer with two coaxial impinging streams compared to a geometrically similar spouted-bed dryer (Tamir and Shalmon, 1988). A volumetric heat transfer coefficient 3–4 times higher than that of geometrically similar spouted-bed and cyclone dryers and around 20 times higher than spray dryers has been reported for an ISD with curvilinear, counter-current configuration.

Both constant and falling drying rate periods were observed in convective drying with impinging streams (without inert media) when internal as well as external resistance to mass flow were present. Because of high heat and mass transfer rates, constant-rate period drying was very short (less than a second). As the duct walls are heated, drying capacity and thermal efficiency may be enhanced by additional conduction of heat.

2.2.8 Infrared Drying

Infrared (IR) heating is widely used in various industries, but despite its potential it has little application in the food sector. There has nevertheless been extensive research on the use of IR heating in food dehydration. The wide IR spectrum is divided into three different segments: near infrared (0.75–1.4 μm), mid-infrared (1.4–3 μm), and far infrared (3.0–1000 μm). In general, the 2.5–7 μm wave band (intersection of mid-and far infrared) is chosen for use in food processing, because components present in food (water, proteins, starches) have high absorbances in this band (Das and Das, 2010). Peak or maximum absorbances for major food constituents have been reported elsewhere (Sandu, 1986):

- 3–4 μm and 6–9 μm for proteins, amino acids, and polypeptides
- 3–4 μm, 6.0 μm, and 9–10 μm for lipids
- 3 μm and 7–10 μm for sugar
- 3, 4.7, 6, and 15.3 μm for liquid water

Reduced drying time, high energy efficiency, good product quality, precise control of the process, reduced space requirement and least environmental hazard are several advantages of the wider use of IR heating and drying in foods (Krishnamurthy et al., 2008). A 50% reduction in drying time and energy density up to 6–10 times higher have been observed for IR drying over convective drying (Nowak and Lewicki, 2004). Typically, energy consumption in IR drying was reduced by 245% compared with drying in hot air at 70°C (Afzal et al., 1999). Prolonged exposure in IR drying caused swelling of the biological materials, leading to disintegration (Das et al., 2004).

Reasonable control of radiation intensity, exposure time, and agitation of the materials can moderate the damage and preserve product quality, therefore IR should be combined with other modes of heating to synergistically enhance overall process efficiency (Dostie et al., 1989). This was observed in a comparison of the thin-layer drying of barley using a combined IR/convection system and a convection system alone (Afzal et al., 1999). Although IR has been successfully applied in recent years to the drying of fruit and vegetable products, both alone and in combination with other heating modes, its use is limited to pilot scale. Successful commercial application of IR in the food-processing and drying sectors with sustained benefits in cost, productivity, energy analysis, and finished product quality remains a challenge, and extensive research is required.

A good example of a recently conceived IR drying technique is refractance window drying, in which a radiating surface-emitting long wavelength is maintained by circulating water near boiling temperature (Raghaviet al., 2018). Heat transfer by radiation and in conjunction with conduction and convection heat transfer accelerates this drying process. This is covered in a separate section below.

2.2.9 Osmotic Drying

In osmotic dehydration (OD) of foods, the food is immersed in a hypertonic solution. Water (moisture) content in the food is reduced, with a consequent increase in its soluble solid content. The concentration gradient of the solute between the solution and cellular fluid is the driving force (osmotic pressure) in the transfer of water (solvent) across the cell wall (semi-permeable membrane). Diffusion of solute from the solution into the food and that of water from food into solution occurs at normal atmospheric pressure or under low pressure (vacuum) (Falade and Igbeka, 2007). Thus, final moisture content in the osmo-dehydrated fruits is reduced, helping to extend their shelf life.

Sucrose and sodium chloride (NaCl) are the most common osmotic agents. Mixed osmotic agents in solutions for dehydration of fruits, vegetables, and meat have also been reported. Osmo-dehydration of apple was studied using an aqueous solution of sucrose with lactose, malto-dextrin, and sodium chloride (Argaiz et al., 1994). The technique was used in dehydration of diced potato with a solution containing a mixture of NaCl and sucrose (Lenart and Flink, 1984). Similarly, NaCl and ethanol in water has been used for osmotic dehydration of sliced carrots (Le Maguer and Biswal, 1984). Other osmotic agents are also used, principally glucose, lactose, malto-dextrin, corn syrup, ethanol, and glycerine (Shi and Xue, 2009).

This osmo-dehydration technique is commonly used for preparation of intermediate-moisture foods and as a pre-drying treatment process. It has obvious advantages over other drying processes: it minimizes heat damage to the product, causes the least discoloration, facilitates good flavor retention, and significantly reduces energy consumption (Khan, 2012). A significant reduction in drying time (around 65%) is reported when osmotic drying is used for the initial phase of the dehydration process for fruits and vegetables (Raoult-Wack et al., 1992). Since removal of water in osmotic drying occurs without any phase change, this system is faster than air drying or freeze-drying.

The rate of exchange of masses (solute and water) in the osmo-dehydration process depends on several factors, including concentration of solute in solution, contact (soaking) time, temperature of the process, mass ratio of solution and food, surface area per unit mass of the food (Shi and Xue, 2009). Since the material (food) cannot be dried by this osmosis process alone, other modes of drying should be used in conjunction during the second phase to achieve good final product quality. These secondary drying processes could be hot-air drying, fluidized-bed drying, vacuum drying or freeze-drying.

Osmotic drying produces changes in the initial composition of the material (food) as well as the final product structure, influencing physical attributes of the product such as color, texture, volume (shrinkage), water-holding capacity and porosity, and sensory attributes, including flavor and taste (Falade and Igbeka, 2007). Even though significant progress has been made in understanding fundamentals of the osmotic drying process, and trials of industrial applications have been carried out, extensive theoretical analysis of the system is required in order to achieve a rational design for large-scale production.

2.3 PULSE COMBUSTION DRYING

Pulse combustion drying consists of intermittent (pulse) combustion of fuel (solid, liquid, or gaseous) in a conventional burner causing intensive pressure, velocity, and (to some extent) temperature waves to be generated in the combustion chamber and reach the dryer through a diffuser (tailpipe). This intermittent combustion technique, which leads to periodic momentum transfer, not only increases productivity, maximizes utilization of the input fuel (combustion intensity up to 10 times; improvement of thermal efficiency up to 40%), and reduces pollutant emissions by up to three times, but also intensifies the rates of heat and mass transfer by 2–5 times and reduces space requirements for the combustion equipment (Mujumdar, 1991; Kudra and Mujumdar, 1995).

The mixture of air and fuel (gas jet or liquid spray) is drawn into the combustion chamber and ignited, the inlet valve for the fuel/air mixture is closed, and the combustion-generated pressure pushes the combustion product through the tailpipe into the process zone (dryer). The pressure in the combustion chamber continues to decrease gradually with the outward momentum (with the discharged flue gas). It reaches a minimum when the fresh fuel/air mixture enters after the valve reopens and the next cycle starts. These combustion cycles occur with specific frequency depending on the design of the pulse combustor and characteristics of the tailpipe.

Typically, pulse combustors operate at frequencies from 20 to 250 Hz. An oscillation of ±10 kPa (pressure of gas) in the combustion chamber produces a variation of gas velocity at the tailpipe exit from 0 to 100 m/s (Keller et al., 1992).

2.4 REFRACTANCE WINDOW (RW) DRYING

In this relatively new drying technology, a modified form of IR drying, food is indirectly heated (Nindo and Tang, 2007). A thin plastic film (Mylar) allows transmission of infrared radiation (3.0–15.3 μm), which is absorbed by moisture present in the food.

The dryer consists of a continuous belt made of thin Mylar sheet floating on the surface of hot water in a shallow trough the same width as the belt. The material to be dried (mainly high viscosity material for liquid foods) is placed on the moving belt (speed: 0.6–3.0 min per minute) in a thin layer (1 mm thick). The water in the trough is maintained at 95–98°C. The temperature of the product remains below 70°C. In addition to heat received from the hot water, the material experiences evaporative cooling and convective heat transfer to the ambient air, causing it to remain at low temperature.

During initial drying, the wet material covering the belt surface receives heat from the water by conduction and radiation. At this point, a passage (window) for transmission of infrared energy is generated over the entire surface due to the minimization of refraction at the belt–material interface (Kudra and Mujumdar, 2009). When sufficient moisture is removed, this infrared window is gradually closed. At this point, heat transfer by conduction from the hot water predominates. One advantage of this process is that the belt material (plastic) is not a good conductor of heat, so there is minimal overheating of the dried material.

At the discharge end of the belt conveyor, the cooling section significantly reduces the temperature of the final product to below the glass transition temperature. This facilitates peeling by avoiding stickiness caused by cohesion and adhesion of the product.

Limited literature is available on the kinetics of refractance window drying. The initial constant-rate stage with thin-layer drying of the feed material is short, followed by a longer falling-rate period (Nindo et al., 2003). Energy consumption is similar to other kinds of drying, i.e., the latent heat of evaporation is the major part of the energy demand. As a rough estimate, the thermal efficiency of RW drying (52–77%) is comparable to that of drum drying, but higher than thermal efficiency of hot-air drying (Nindo et al., 2003; Nindo and Tang, 2007).

Refractance window drying is suitable for concentration and dehydration of suspensions, pastes, and solutions (Anonymous, 2007). Solids containing high moisture can also be dried. For easy and uniform spreading of the materials on the conveyor belt, chopped solid materials or liquid of the appropriate consistency should be used. This technology is quite suitable for heat-sensitive material as the drying is carried out at low product temperature under normal atmospheric pressure, and the drying time is relatively short. These conditions are quite favorable for drying of fermented milk products, such as yogurt, curd, and kefir, without any apparent loss of microbial activity. Thus, their beneficial probiotic effect is preserved. The reconstitution properties of dried granular products are also favorably faster.

Some authors (Abonyi et al., 2002; Nindo et al., 2003) have observed that this process inhibits non-enzymatic browning, and retains most of the quality attributes and nutritional components in the final product. The final quality of RW-dried products is quite comparable to that of freeze-dried products, and processing costs are lower. The constructional features of the drying system are much simpler and the operating cost is affordable.

2.5 SLUSH DRYING

An ice–liquid mixture is called slush. Slush drying and freeze-concentration of materials are quite similar processes. However, there is an exception. Separation of ice from the supernatant liquid is carried out by sublimation together with evaporation of part of the water from the concentrated liquid. The high dissolved solids content in the unfrozen liquid favors good volatiles retention. There is no need to maintain ultra-low temperatures as in the case of freeze-drying.

Slush drying is essentially a batch process. The liquid (fruit juice or similar material) undergoing concentration is first cooled down to a specific temperature between the freezing point of the pure solvent and the eutectic temperature of the solution or the slurry. The resultant semisolid material is then placed in a vacuum chamber with pressure around 1 mm Hg. Heat is supplied by an internal heater for sublimation of ice crystals. The vacuum pump is used to evacuate both water vapor and volatile compounds, followed by cooling and condensation of these materials using a vapor trap. Depending upon the initial level of dissolved solids in the fruit juice, stable and coarse sugar-rich foam for selective retention of aroma is formed at the surface of the bed. In fact, the sugars present in the foam help to lock the volatiles. It could be concluded that slush drying is better than air or vacuum drying when it comes to aroma retention.

Slush drying has numerous advantages over freeze-drying and freeze-concentration. Since sublimation of vapor occurs from the frozen-zone temperature, fruit juices and other sugar-containing substances have lower drying rates. In slush drying, the vapor pressure of water at high operating temperature is higher, so drying rates are comparatively higher. The cost of refrigeration in slush drying is low, as there is no need for a low-temperature condenser. There is less degradation of product quality and higher retention of aroma with slush drying than with freeze concentration.

2.6 ATMOSPHERE FREEZE-DRYING

Vacuum freeze-drying of high-value products such as enzymes, nutraceuticals, pharmaceuticals, and important products for biotechnology uses is a unique but expensive process. Cost reduction can be achieved, however, with some modifications of the process: eliminating condensers; enhancing contact between the phases responsible for mass and heat transfer; and removing water at normal atmospheric pressure.

In this process, a suitable solid adsorbent comes into contact with the frozen product. The sublimed vapor is adsorbed by the adsorbent, which acts as an external condenser. Both fixed-bed and fluidized-bed drying of adsorbent particles showed good results. In fixed-bed drying, layers of frozen material are alternated with layers of adsorbent material and a low-pressure gas flows through the layers. The sorbent adsorbs moisture from this air (exothermic process), and the warm air performs the sublimation. Due to the low heat and mass transfer, the drying process is very slow (Kudra and Mujumdar, 2002).

In the fluidized bed, the drying process is faster, with random collisions between the frozen particles and the adsorbent. These adsorbent particles act as the medium for heat transfer as well as the transport system for water vapor. In this well-mixed fluidized bed, the adsorbent particles have different water moisture saturation levels. Moisture transfer occurs not only between adsorbent and frozen product, but also among the adsorbent particles in the bulk.

Thus a continuous process of freeze-drying at normal atmospheric pressure is easily achievable. The main limitations of the atmosphere freeze-drying system are extended processing time and difficulty separating adsorbents from the dried product. The use of adsorbent material that is compatible with the dried product is therefore suggested. Drying of frozen potato using dry starch as adsorbent (Wolff and Gibert, 1990) and frozen shrimp using wheat bran as adsorbent (Donsi et al., 2000) have been reported.

2.7 SUPERHEATED STEAM DRYING

Superheated steam drying has been applied successfully to dry several food products. Superheated steam (SS), which acts as both heating and drying medium, is generated by heating saturated steam to increase its temperature and specific enthalpy at constant pressure. On passing through the target wet materials, the SS transfers heat and picks up the evaporated water. The moisture content of the wet material is gradually reduced with the flow of the SS. The exhaust steam from the drying stage may remain above the condensation temperature at the prevailing pressure but its specific enthalpy becomes low. This can be utilized or recycled in the same or a separate process to minimize energy consumption.

High energy savings (50–80%) are observed for drying with superheated steam over conventional hot-air drying. The SS drying process consists of three drying stages, as in conventional drying: initial heating period, constant-rate period and falling-rate period (Pronyk, 2007). Energy analysis shows 0.2–2.5 MJ are required per kg of water evaporated for SS drying, compared with 2.9–4.1 MJ per kg in hot-air drying systems (Chryat et al., 2019). Several authors describe the principles, advantages, disadvantages, and wide applications of drying techniques using superheated steam (Kumar and Mujumdar, 1990; Mujumdar, 1992; Kudra and Mujumdar, 2009; Karimi, 2010).

Superheated steam drying does not use oxygen so there is no possibility of oxidation or adverse reactions and no risk of explosion or fire hazard. Under some specific conditions, SS drying showed higher drying rates in both constant and falling-rate periods. Reports are available of SS drying of food products including green tea, sugar beet pulp, shrimp, potato chips, fries, nuts, wheat, flour, vegetables, noodles, herbs, and spices (Pronyk et al., 2005; Prachayawarakorn et al., 2002; van Deventer and Heijmans, 2001; Tang et al., 2000; Urbaniec and Malczewski, 1997).

Despite its advantages, SS drying of foods has not been widely accepted for industrial use because of a number of inherent drawbacks. These are high capital costs, expensive energy recovery process, larger space requirement for installation of equipment, cleaning of steam and, most importantly, unfavorable high temperature (Wimmerstedt and Hager, 1996). There have been some studies of sub-atmospheric pressure superheated steam low-temperature drying of food.

The choice of dryer is usually determined by cost and energy source (mostly electricity and combustion fuels). High energy efficiency can be achieved using vapor-recompression technology in superheated steam drying. Although there are some niche markets for superheated steam dryers, more robust engineering design is needed before this technology can be used practically for drying of foods.

2.8 SOLAR DRYING

Since the dawn of civilization, people have practiced sun drying in open space to preserve various agricultural or forest materials, including fruits, fish, meat, plants, seeds, and wood. Apart from the unlimited availability of this renewable energy source, there are several technological, quality, and economic limitations to its use in commercial drying of agricultural products. There have been many attempts to harness solar energy for drying systems, including the development in recent decades of solar dryers. Periodic change of solar insolation (intensity or energy flux density) during the hours of sunshine requires a large collection area (solar flat-plate collector) or conversion of the energy to a smaller area (tracking parabolic collector). Heat storage devices or an auxiliary energy source are also used with these solar dryers.

The solar radiation level is quite favorable for drying as this operation requires less energy than other food preservation or processing operations. The investment costs, as well as economic utilization of this energy for drying, need several coordinations: viz.. between the energy demand and the radiation characteristics of solar energy. The flat-plate collector is cheaper and is suitable for drying systems that require the heated medium (flowing air) temperature to be in the range of 60–70°C. Dryers coupled with flat-plate collectors are suitable for drying of common agricultural products. Expensive concentrating collectors should be used where a high-temperature heating medium is required.

The main component in a solar dryer is the heat-absorbing surface (absorber). It absorbs the solar spectrum of radiation, and its temperature is raised by the conversion of solar energy into thermal energy. This energy is then transferred to the fluid (air or water) flowing in contact with the surface of the absorber. The level of radiation heat transfer between the incident solar energy and the absorber surface is governed by specific radiation properties of the absorber (absorptivity and emissivity), surface geometry, view factor, area, and the secondary heat loss of the absorbed energy to the surroundings. The efficiency of the heat collection and the simultaneous transmission of the absorbed energy to the working fluid depend on the design features of the collector. In addition to the heat-absorbing surface, a glazing surface (transparent glass or polymer sheet) is laid over the absorber, and a gap is maintained between them to act as a flow channel for the air. The glazing surface should be transparent to solar radiation (0.25–2.5 μm) but opaque to thermal radiation (IR and far IR) so that solar radiation can be trapped by the absorber. It also prevents heat loss by emission from the absorber to the surroundings in what is known as the greenhouse effect.

Loss of convection heat from the top glazing surface due to the outside air current is reduced with a second layer of glazing. In a double-glazing solar flat-plate collector, a stagnant air gap (air insulation) is maintained between the glazing sheets. In this arrangement, some attenuation of the incoming solar radiation on the absorber surface occurs, but its level is small. A layer of insulation at the back of the absorber minimizes heat loss from the bottom surface. The surface of the absorber (or the whole collector) faces south to receive higher solar insolation (directional effect of radiation heat transfer) throughout the hours of sunshine.

Solar dryers are classified according to their mode of receiving heat (flat-plate or concentrating collectors) or their mode of operation in deriving heat from the heated surface. The discussion here is confined to solar dryers with flat-plate collectors, which are well suited to low-temperature drying.

The two basic operating modes are natural-convection and forced-convection solar dryers. In the natural-convection system, continuous air movements established as a result of temperature rise (difference in density) of the air (buoyancy effect), without any external mechanical or electrical device. Thus, the system is self-operated. The moderate airflow rate generated in the natural-convection system is adequate for the drying process but the rate is slow. This type of solar-drying system works well in tropical and sub-tropical areas with long hours of sunshine and a higher level of solar insolation (solar radiation flux). The forced-convection dryers employ a fan or blower to maintain a higher velocity of hot air in the system. This arrangement facilitates faster drying of the wet materials.

A further classification is direct-mode and indirect-mode solar dryers. In direct-mode solar dryers, the materials to be dried are placed inside and the energy of the sun falls directly on them after it has passed through the glazing surface. In other words, solar radiation is absorbed by the materials along with the exposed part of the absorbing surface. In the case of indirect-mode solar dryers, solar energy is not directly received by the materials inside the dryer. Air is heated by passing through the flow channel in the solar flat-plate collector. This heated air then flows into the drying section to remove moisture from the wet material. Although these direct and indirect mechanisms are applied in both natural- and forced-convection solar dryers, the indirect-mode operation is more common for forced-convection dryers.

The solar cabinet dryer is the simplest form of direct-mode solar dryer (Lof, 1962). The system consists of a small rectangular box (length:width ratio usually 3:1) made of insulating material (generally wooden planks) with a perforated bottom. All the interior surfaces of the cabinet are painted black for high absorption of incident solar energy. A transparent cover (single- or double-layered) is placed over the box, which is tilted to face the south. This arrangement facilitates reduction of reflective losses from the glazing surface and the absorbing surface receives more transmission of the solar radiation. The wet materials are kept inside the cabinet on perforated metal trays. The air enters from the bottom holes, while heated air escapes through ventilation holes by natural convection. Thus, constant air circulation is established. Reduction of drying time of half to one-third has been reported in a cabinet solar dryer compared to sun drying on an open surface. Performance can be improved using a draft chimney with the system, and these dryers are considered better for small-scale operations in tropical and sub-tropical regions. The tent dryer is another version of the cabinet dryer. It consists of a triangular rigid metal framework covered with a thin transparent sheet and south-oriented front wall. The back wall is covered with a black sheet. The material is spread inside the tent on perforated trays.

The capacity of cabinet dryers is small, and the drying rate is slow owing to the low airflow rate (low-pressure differential between inlet and outlet ports). Exploiting the chimney effect (vertical flow of hot air) by using a self-drafted solar dryer increases both air circulation rate and capacity. The tilted collector facing south heats the flowing air. This hot air is allowed to pass through a dryer containing a stack of perforated shelves loaded with materials to be dried. The dryer is attached to a vertical collector (facing south) that facilitates heating and enables the air to escape from the top due to the pressure differential (chimney effect), establishing a constant high airflow rate. The capacity of this dryer can be increased further by increasing the heat collection area and the magnitude of the air draft via the chimney height. Solar dryers with a heat storage option are available, and those can be used for extended drying periods (night-time/cloud conditions).

Drying time in natural-convection solar dryers depends on the initial moisture content of the materials, material depth in the bed, solar insolation level and other meteorological parameters (temperature and humidity of the ambient air). The drying rate is inversely proportional to the bed thickness. Air circulation by natural convection is greatly hindered in the packed-bed configuration. In some modified solar dryer designs, efficiency of the drying system has been improved using a wind-driven fan at the exhaust end of the chimney (Kilkis, 1981).

In forced-convection solar dryers, a larger mass of materials is dried within a shorter time with high circulation of air through the wet material using a power-operated blower. Drying of grains in situ using forced-convection bin-type solar dryers is quite common. In this system, where a false or perforated floor (open area ≥15%) is installed, slow drying at low temperature is preferred. Warm air (with 5–10°C rise in temperature) can typically be conveniently generated with an integrated array of forced-convection solar air-heating systems (Das and Chakraverty, 2003).

In a typical vertical wall circular bin solar grain dryer, half of the south-facing wall is converted into a flat-plate collector. This part of the wall is painted black, and a transparent UV-stabilized plastic film/sheet is laid on it, maintaining a gap between the wall and the sheet, which becomes an airflow channel. During the winter months, the performance of the solar bin dryer is higher because of the declination of the sun and the incidence angle of the solar radiation on the vertical surface which causes a higher level of solar insolation on the wall. Under these conditions, a reduction in crop drying time of as much as 50–75% has been observed. The rise in air temperature varied between 5 and 12°C (Sodha et al., 1987).

A large collector area is required for commercial-scale drying of grain using a solar air-heating system, and this is achieved simply by scaling up from a small solar collector unit. The thermal efficiency of an air-heating system using an assembly of small collectors is reduced due to large thermal losses from the side walls (Chakraverty and Das, 1990). However, this thermal loss could be mitigated with an integrated array of collectors in which airflow channels are arranged internally (Das and Chakraverty, 2003).

Performance of the solar air-heating system over the course of a year is dependent on the average level of radiation flux received on the collector surface. The optimum tilt of the collector is important for receiving high solar radiation flux throughout the year. This, in turn, depends on the location's latitude and longitude.

As solar radiation is intermittent in nature, coupling an auxiliary air-heating system with a solar air-heating system is a feasible option for continuous drying during non-availability periods. This auxiliary energy may be derived from burning of conventional or non-conventional fuels (oil/gas or biomass) or electricity. Storing solar energy using rocks or similar low thermal diffusivity materials is another low-cost option. Storing energy using the phase-change principle is another option, but the cost may be prohibitive.

2.9 HYBRID DRYING

Generally, a single type of dehydration process is not optimally and exclusively suitable for all kinds of food materials in terms of energy, cost, and final product quality. About 85% of industrial and commercial dryers are the convective type which demand a high amount of specific energy (Kudra, 2009). Hybrid drying techniques could be a feasible way of overcoming these limitations. In a hybrid drying system, various combinations of drying techniques are applied in sequence to eliminate undesirable impacts with regard to energy, cost, and quality. Combining different energy sources for drying is quite beneficial for both energy consumption and product quality (Grabowski et al., 2003).

There are three main categories of hybrid drying: (1) multistage dryer systems; (2) multi-processing dryer systems; and (3) systems that use combined modes of heat transfer (Kudra and Mujumdar, 2009). Higher drying rates, good-quality dried products, effective waste-heat utilization, and energy saving have been reported with hybrid drying systems. Typical examples of hybrid drying systems are hot air-assisted, solar-assisted, microwave-assisted, and infrared-assisted drying systems. Reductions of 20–50% in energy consumption over conventional drying systems have been reported with multistage hybrid drying (Chou and Chua, 2001; Vega-Mercado et al., 2001; Erle, 2005; Soysal et al., 2006). A comparative study on the final quality of dehydrated carrot and pumpkin samples using hybrid drying (microwave-assisted freeze-drying) and freeze-drying alone showed an insignificant difference (Donsi et al., 1998; Kumar et al., 2001). A 50% saving in total energy consumption, as well as a reduction in drying time, has also been observed. Drying of shrimp using combined drying systems, such as superheated steam drying-assisted heat-pump drying, showed significant improvement in the quality of the final dried product (Namsaguan et al., 2004). Extensive work is required on the challenges of control, optimization, and scaling up of hybrid systems.

2.10 CONCLUSIONS

Drying is the oldest but still the most essential and effective industrial operation in all major sectors that deal with agricultural, chemical, pharmaceutical, textile, leather, and paper products. Dealing with moisture removal by supplying heat is by definition not a simple unit operation. The retention of components while maintaining various quality factors in the finished products remains a great challenge. Against a backdrop of a continuing energy crisis and consumer demand for quality, several path-breaking initiatives have emerged as future drying technologies. The task of designing an effective and energy-efficient drying system is made more complex by factors such as the diverse characteristics (physical and chemical) of both wet and final dried materials, different modes of energy supply, the binding properties of water with food components, cost, energy recycling, waste-heat utilization, and environmental constraints. Indeed, advanced research in food drying needs an inter- and multi-disciplinary approach. While the commercial success of these newly developed drying technologies is not yet apparent, their potential application in the near future cannot be discounted.

REFERENCES

Abonyi, B.I., H. Feng, J. Tang, C.G. Edwards, D.S. Mattinson, and J.K. Fellman. 2002. Quality retention in strawberry and carrot purees dried with Refractance Window system. *Journal of Food Science*, 67: 1051–1056.

Afzal, T.M., T. Abe, and Y. Hikida. 1999. Energy and quality aspects during combined FIR convection drying of barley. *Journal of Food Engineering*, 42: 177–182.

Anonymous. 2007. *Dryer: Refractance Window®*. Tacoma, WA: MCD Technologies, Inc.

Argaiz, A., A. Lopez-Malo, E. Paou, and J. Welti. 1994. Osmotic dehydration of papaya with corn syrup solids. *Drying Technology*, 12: 1709–1725.

Arzhavitina, A. and H. Steckel. 2010. Foams for pharmaceutical and cosmetic application. *International Journal of Pharmacology*, 394: 3–7.

Baker, G.J. 1997. *Industrial Drying of Foods*. London: Chapman and Hall.

Barbosa Canovas, G.V. and H. Vegas-Mercanto. 1996. *Dehydration of Foods*. New York: International Thompson Publishing.

Chakraverty, A. and S.K. Das. 1990. Performance of an integrated array of solar collector modules for different air flow rates and flow organisations. *Energy Conversion and Management*, 30: 29.

Chou, S.K. and K.J. Chua. 2001. New hybrid drying technologies for heat sensitive foodstuffs. *Trends in Food Science and Technology*, 12: 359–369.

Chryat, Y., M. Esteban-Decloux, C. Labarde, and H. Romdhana. 2019. A concept and industrial testing of a superheated steam rotary dryer demonstrator: Cocurrent-triple pass design. *Drying Technology*, 37(4): 468–474.

Chua, K.J., S.K. Chou, J.C. Ho, and M.N.A. Hawlader. 2002. Heat pump drying: Recent developments and future trends. *Drying Technology*, 20(8): 1579–1610.

Das, I. and S.K. Das. 2010. Emitters and infrared heating system design. In: Pan, Z. and Atungulu, G.G. (eds.), *Infrared Heating for Food and Agricultural Processing*. Florida: CRC Press, pp. 57–88.

Das, I., S.K. Das, and S. Bal. 2004. Drying performance of a batch type vibration-aided infrared dryer. *Journal of Food Engineering*, 4: 129–133.

Das, S.K. and A. Chakraverty. 2003. Grain drying systems. In: Chakraverty, A., Mujumdar, A.S., Raghavan, G.S.V., and Ramaswamy, H.S. (eds.), *Handbook of Postharvest Technology*. New York: Marcel Dekker, pp. 139–166.

Daschner, F., M. Kent, R. Knochel, and U.K. Berger. 2000. Optimization of the microwave determination of water in foods using principal component analysis. In: *Proceedings of the 17th IEEE Instrumentation and Measurement Technology Conference*, 1: 12–16.

Dincer, I. and C. Zamfirescu. 2016. *Drying Phenomena: Theory and Applications*. New York: John Wiley.

Donsi, G., G. Ferrari, and P.D. Matteo. 2000. *Atmospheric and vacuum freeze-drying kinetics of shrimps. Proceedings of 12th International Drying Symposium (IDS-2000)*, Amsterdam, The Netherlands, Paper no 279.

Donsi, G., G. Ferrari, R. Nigro, and P.D. Maltero. 1998. Combination of mild dehydration and freeze-drying processes to obtain high quality dried vegetables and fruits. *Transactions of the Institution of Chemical Engineers*, 76: 181–187.

Dostie, M., J.N. Seguin, D. Maure, Q.A. Ton-That, and R. Chatingy. 1989. Preliminary measurements on the drying of thick porous materials by combinations of intermittent infrared and continuous convection heating. In: Mujumdar, A.S. and Roques, M.A. (eds.), *Drying '89*. New York: Hemisphere Publications.

Erle, U. 2005. Drying using microwave processing. In: Schubert, H. and Regier, M. (eds.), *The Microwave Processing of Foods*. Cambridge: Woodhead Publications, pp. 142–152.

Falade, K.O. and J.C. Igbeka. 2007. Osmotic dehydration of tropical fruits and vegetables. *Food Reviews International*, 23(4): 373–405.

Fayose, F. and Huan, Z. 2016. Heat pump drying of fruits and vegetables: Principles and potentials for sub-Saharan Africa. *Annals of the Faculty of Engineering Hunedoara-International Journal of Engineering*, 14(1): 207–214.

Febrianto, A., S. Kumalaningsih, and A.W. Aswari. 2012. Process engineering of drying milk powder with foam-mat drying method. A study on the effect of the concentration and types of filler. *Journal of Basic and Applied Science and Research*, 2(4): 3588–3591.

Felke, K., T. Pfeiffer, and P. Eisner. 2009. Neues Verfahren zur schnellen und schonenden Erhitzung von verpackten Lebensmitteln: Hochfrequenzerhitzung im Wasserbad. *Chemie Ingenieur Technik*, 81(11): 1815–1821.

Grabowski, S., M. Marcotte, and H.S. Ramaswamy. 2003. Drying of fruits, vegetables, and spices. In: Chakraverty, A., Mujumdar, A.S., Raghavan, G.S.V., and Ramaswamy, H.S. (eds.), *Handbook of Postharvest Technology: Cereals, Fruits, Vegetables, Tea, and Spices*. New York: Marcel Dekker, pp. 653–695.

Grbavcic, Z.B., Z.L. Arsenijevic, and F.K. Zdanski. 1998. *Drying of suspensions in fluidised bed of inert particles*. In: Akritidis, C.B., Marinos-Kouris, D., and Saravakos, G.D. (eds.), *Drying '98: Proc. 11th Intl. Drying Symposium*, Halkidiki, Greece, pp. 2090–2097.

Hardy, Z. and V.A. Jideani. 2017. Foam-mat drying technology: A review. *Critical Reviews in Food Science and Nutrition*, 57(12): 2560–2572.

Hawlader, M.N.A. and K.A. Jahangeer. 2012. Heat pump design for food processing. In: Ahmed, J. and Rahman, M.S. (eds.), *Handbook of Food Process Design*. Oxford: Wiley-Blackwell, pp. 578–620.

Indrawati, L., Wang, Z., Narsimhan, G., and J. Gonzalez. 2008. Effect of processing parameters on foam formation using a continuous system with a mechanical whipper. *Journal of Food Engineering*, 88: 65–69.

Jiang, H., Liu, Z., and Wang, S. 2017. Microwave processing: Effects and impacts on food components. *Critical Reviews in Food Science and Nutrition*, 58(14): 2476–2489.

Jiang, H., M. Zhang, and Mujumdar, A.S. 2010. Microwave freeze-drying characteristics of banana crisps. *Drying Technology*, 28: 1377–1384.

Kadam, D.M., R.T. Patil, and Kaushik, P. 2010. Foam mat drying of fruit and vegetable products. In: Jangam, S.V., Law, C.L., and Mujumdar, A.S. (eds.), *Drying of Foods, Vegetables and Fruits – Vol. 1*, pp. 111–124. ISBN: 978-981-08-6759-1.

Karimi, F. 2010. Applications of superheated steam for the drying of food products. *International Agrophysics*, 24: 195–204.

Keller, J.O., German, R.S., and R.W. Ozer. 1992. Fundamentals of enhanced scalar transport in strongly oscillating and/or resonant flow fields as created by pulse combustion. In: Mujumdar, A. S. (ed.), *Drying '92*. Amsterdam: Elsevier Science Publishers, pp. 161–180.

Khan, M.R. 2012. Osmotic dehydration technique for fruits preservation: A review. *Pakistan Journal of Food Science*, 22: 71–85.

Kilkis, B. 1981. *Solar Energy Assisted Crop and Fruit Drying Systems: Theory and Applications, Energy Conservation and Use of Renewable Energies in Bio-industries*. Oxford: Pergamon Press, pp. 307–333.

Krishnamurthy, K., H.K. Khurana, S. Jun, J. Irudayaraj, and A. Demirci. 2008. Infrared heating in food processing: An overview. *Comprehensive Reviews in Food Science and Food Safety*, 7: 2–13.

Kudra, T. 2009. Energy aspects in food dehydration. In: Ratti, C. (ed.), *Advances in Food Dehydration*. Boca Raton: CRC Press, pp. 423–443.

Kudra, T. and Mujumdar, A.S. 1995. Special drying techniques and novel dryers. In: Mujumdar, A.S. (ed.), *Handbook of Industrial Drying*, 2nd edn. New York: Marcel Dekker, pp. 1087–1149.

Kudra, T. and Mujumdar, A.S. 2002. *Advanced Drying Technologies*. New York: Marcel Dekker. doi: 10.1021/op010102p

Kudra, T. and Mujumdar, A.S. 2009. *Advanced Drying Technology*, 2nd edn. Florida: CRC Press.

Kudra, T. and Ratti, C. 2006. Foam-mat drying: Energy and cost analyses. *Canadian Biosystem Engineering*, 48: 3.27–3.29.

Kumar, H.S.D., K. Radhakrishna, P.K. Nagaraju, and D.V. Rao. 2001. Effect of combination drying on physico-chemical characteristics of carrot and pumpkin. *Journal of Food Processing and Preservation*, 25: 447–460.

Kumar, P. and Mujumdar, A.S. 1990. Superheated steam drying—A review. In: Mujumdar, A.S. (ed.), *Drying of Solids*. India: Sarita, Meerut.

Kutsakova, V.E. 2007. Drying of liquid and pasty products in a modified spouted bed of inert particles. *Drying Technology*: 2343–2350.

Le Maguer, M. and R.N. Biswal. 1984. Multicomponent diffusion and vapour–liquid equilibria of dilute organic components in aqueous sugar solutions. *American Institute of Chemical Engineering Journal*, 18: 513–519.

Lenart, A. and J.M. Flink. 1984. Osmotic concentration of potato. I. Criteria for the end-point of the osmosis process. *Journal of Food Technology*, 19: 45–63.

Levy, A. and Borde, I. 2015. Pneumatic and flash drying. In: Mujumdar, A.S. (ed.), *Handbook of Industrial Drying*, 4th edn. Florida: CRC Press, pp. 381–391.

Lof, G.O.G. 1962. Recent investigations in the use of solar energy for drying of solids. *Solar Energy*, 6: 122.

Micu, A. 2016. Meet Chuño, a space worthy food that the Incas made eight centuries ago. https://www.zmescience.com/medicine/nutrition-medicine/chuno-taters, accessed on 17-01-2020.

Mortezapour, H., B. Ghobadian, S. Minaei, and M.H. Khoshtaghaza. 2012. Saffron drying with a heat pump-assisted hybrid photovoltaic-thermal solar dryer. *Drying technology: An International Journal*, 30(6): 560–566.

Moses, J.A., T. Nórton, K. Alagusundarm, and B.K. Tiwari. 2014. Novel drying techniques for food industry. *Food Engineering Reviews*, 6(3): 43–55.

Mujumdar, A.S. 1991. Drying technologies of the future. *Drying Technology*, 9(2): 325–347.

Mujumdar, A.S. 1992. Superheated steam drying of paper: Principles, status and potential. In: Mujumdar, A.S. (ed.), *Drying of Solids*. New York: International Science Publisher, pp. 208–220.

Mujumdar, A.S. 2000. Dryers for particulate solids, slurries and sheet-form materials. In: Devahastin, S. (ed.), *Mujumdar's Practical Guide to Industrial Drying*. Montreal: Exergex Corp., pp. 37–71.

Mujumdar, A.S. (ed.). 2015. *Handbook of Industrial Drying*, 4th edn. New York: CRC Press, pp. 3–29. doi: 10.1201/b17208.

Mujumdar, A.S. and A.S. Menon. 1995. Drying of solids: Principles, classification, and selection of dryers. In: Mujumdar, A.S. (ed.), *Handbook of Industrial Drying*, Vol. 1, 2nd edn. New York: Marcel Dekkerpp., pp. 1–39.

Namsanguan, Y., Tia, W., S. Devahastin, and S. Soponronnarit. 2004. Drying kinetics and quality of shrimp undergoing different two-stage drying processes. *Drying Technology*, 22: 759–778.

Nindo, C.I., H. Feng, G.Q. Shen, J. Tang, and D.H. Kang. 2003. Energy utilization and microbial reduction in a new film drying system. *Journal of Food Processing and Preservation*, 27(2): 117–136.

Nindo, C.I. and J. Tang. 2007. Refractance Window dehydration technology: A novel contact drying method. *Drying Technology*, 25(1): 37–48.

Nowak, D. and P.P. Lewicki. 2004. Infrared drying of apple slices. *Innovative Food Science and Emerging Technologies*, 5: 353–360.

Perera, C.O. and M. Rahman. 1997. Heat pump dehumidifier drying of food, *Trends in Food Science and Technology*, 8(3): 75–79.

Prachayawarakorn, S., Soponronnarit, S., S. Wetchacama, and D. Jaisut. 2002. Desorption isotherms and drying characteristics of shrimp in superheated steam and hot air. *Drying Technology*, 20: 669–684.

Pronyk, C. 2007. Effect of superheated steam processing on the drying kinetics and textural properties of instant Asian noodles. PhD Thesis, University of Manitoba, Winnipeg, Canada.

Pronyk, C., S. Cenkowski, and W.E. Muir. 2005. *Current status of superheated steam drying and processing. Proceedings of the 3rd Inter-American Drying Conference*, 21–23 Aug, Montreal, Canada.

Raghavi, L.M., J.A. Moses, and C. Anandharamakrishnan. 2018. Refractance window drying of foods: A review. *Journal of Food Engineering*, 222: 267–275.

Rajkumar, P., R. Kailappan, R. Viswanathan, and G.S.V. Raghavan. 2007. Drying characteristics of foamed Alphonso mango pulp in continuous type foam-mat dryer. *Journal of Food Engineering*, 79: 1452–1453.

Raoult-Wack, A.L., S. Guilbert, M. Le Maguer, and G. Rios. 1992. Simultaneous water and solute transport in shrinking media: Part 1: Application to dewatering and impregnation soaking process analysis (osmotic dehydration). *Drying Technology*, 9: 589–612.

Ratti, C. 2009. *Advances in Food Dehydration*. Boca Raton: Taylor and Francis Group, CRC Press.

Rizvi, S.S.H. 2005. Thermodynamic properties of foods in dehydration. In: Rao, M.A., Rizvi, S.S.H., and Datta, A.K. (eds.), *Engineering Properties of Foods*, 3rd edn. New York: Taylor and Francis, pp. 259–346.

Rysin, A.P. and A.S. Ginzburg. 1992. Theory and technology of food product drying in fluidized vibrated bed. In: Mujumdar, A.S. (ed.), *Drying of Solids*. New York: International Science Publisher, pp. 86–124.

Sabarez, H. 2016. Drying of food materials. In: *Reference Modules in Food Science*. doi: 10.1016/B978-0-08-100596-5.03416-8.

Sandu, C. 1986. Infrared radiative drying in food engineering; a process analysis. *Biotechnology Progress*, 2(3): 109–119.

Saravacos, G.D. and A.E. Kostarapoulos. 2002. *Handbook of Food Processing Equipments*. Cham: Springer Science, pp. 331–377.

Shi, J. and J.S. Xue. 2009. Application and development of osmotic dehydration technology in food processing. In: Ratti, C. (ed.), *Advances in Food Dehydration*. Boca Raton: CRC Press, pp. 187–208.

Shi, Q., C.H. Xue, Y. Zhao, Z.J. Li, and X.Y. Wang. 2008. Drying characteristics of horse mackerel (Trachurus japonicas) dried in a heat pump dehumidifier. *Journal of Food Engineering*, 84(1): 12–20.

Sodha, M.S., N.K. Bansal, A. Kumar, P.K. Bansal, and M.A.S. Malik. 1987. *Solar Crop Drying*. Vol. 1. Boca Raton: CRC Press.

Soysal, Y., S. Oztekin, and O. Eren. 2006. Microwave drying of parsley modelling, kinetics and energy aspects. *Biosystem Engineering*, 93: 403–413.

Strumillo, C. and T. Kudra. 1986. *Drying: Principles, Applications and Design*. New York: Gordon and Breach.

Strumillo, C., A. Markowski, and W. Kaminski. 1983. Modern developments in drying of paste materials. In: Mujumdar, A.S. (ed.), *Advances in Drying*. Vol. 2. Amsterdam: Hemisphere Publishing Corp., pp. 193–232.

Tamir, A. and B. Shalmon. 1988. Scale-up of two-impinging streams (TIS) reactors. *Industrial Engineering Chemical Research*, 27: 238–242.

Tang, Z., S. Cenkowski, and W.E. Muir. 2000. Dehydration of sugar-beet pulp in superheated steam and hot air. *Transaction of American Society of Agricultural Engineers*, 43: 685–689.

Urbaniec, K. and J. Malczewski. 1997. Experimental investigations of beet pulp drying in superheated steam under pressure. *Drying Technology*, 15(6–8): 2005–2013.

van Arsdel, W.B. and M.J. Copley. 1963. *Food Dehydration, Vol 1: Principles*. Westport: Avi Publishing.

van Deventer, H.C. and M.H. Heijmans. 2001. Drying with superheated steam. *Drying Technology*, 19(8): 2033–2045.

Vega-Mercado, H., M.M. Gongora-Nieto, and G.V. Barbosa-Canovas. 2001. Advances in dehydration of foods. *Journal of Food Engineering*, 49: 271–289.

Wang, Y., L. Zhang, J. Johnson, M. Gao, J. Tang, J.R. Powers, and S. Wang. 2014. Developing hot air-assisted radio frequency drying for in-shell macadamia nuts. *Food Bioprocess Technology*, 7(1): 278–288.

Wimmerstedt, R. and J. Hager. 1996. Steam drying–modelling and applications. *Drying Technology*, 14(5): 1099–1119.

Wolff, E. and H. Gibert. 1990. Atmospheric freeze-drying. Part 1: Design, experimental investigation and energy-saving advantages. *Drying Technology*, 8(2): 385–404.

3 Applications of Evaporation in Food Processing

Saptashish Deb
Department of Food Engineering and Technology, Sant Longowal Institute of Engineering & Technology, Sangrur, Punjab, India

Umexi Rani
Department of Food Engineering and Technology, Sant Longowal Institute of Engineering & Technology, Sangrur, Punjab, India

Santanu Malakar
Department of Food Engineering, National Institute of Food Technology Entrepreneurship and Management, Sonepat, Haryana, India

CONTENTS

3.1 INTRODUCTION

Exponential population growth is causing grave concern to those involved in supplying basic food requirements. Across the world, food scientists and researchers are working to protect food quality and reduce losses to balance supply and demand. The food-processing sector is attracting a great deal of attention among researchers, food technologists, and food industries to ensure the demand is fulfilled with maximum profits and minimal losses (Sunderland, 2011).

In food processing, the moisture level or presence of water is a key factor in preservation of food. Removing water is a viable method of saving foodstuffs from spoilage, reducing transportation costs, and increasing shelf life (Al-Muhtaseb et al., 2002). Various methods have been developed to remove water from food, including drying or dehydration, evaporation, filtration, and centrifugation. Drying and evaporation are the two principal ways of removing water from food by applying heat to separate dissolved solute from a solution (Hayashi, 1989), whereas in techniques like filtration and centrifugation solids are not in a state of solution, only suspended in the solvent and able to be physically separated. Phase change is not required in this kind of separation process (Dechow, 1989). Drying is used to remove water from solid food and the final product is dry solid. Evaporation applies to liquid food in which the water present reaches boiling point and escapes into the surrounding environment. Another distinction is that drying is a simultaneous process of heat and mass transfer, whereas evaporation is a purely heat transfer process (Kerr, 2013; Rao, 2016).

Evaporation comes into the category of physical separation processes. In the case of liquid food, water is separated from the product to make a more concentrated product (Seaderand Henley, 1998). Removal of volatile components from a solution is also a form of evaporation. A variety of evaporation techniques are used: in some cases, concentrated liquid is separated by evaporating excess water to reduce the bulk and make transportation easier, to improve storage and to increase product shelf life by protecting it from microbiological spoilage due to moisture. Fruit juices, jams, jellies, and sugar crystals are examples where the unwanted liquid is evaporated to obtain the desired product (Somogyiand Luh, 1986). In some instances evaporation is used to obtain the desired liquid and the concentrated liquid is rejected, for example in water purification, where water mixed with minerals and solvents is evaporated followed by condensation to obtain the pure water, and the concentrated mixture of solvents and minerals is rejected (Rao, 2016). However, in most cases evaporation is used to concentrate the desired liquid by evaporating the undesired liquid.

The simplest form of evaporation is atmospheric evaporation. In this type of evaporation liquid is kept in an open container, and when it is heated, the vapor escapes from the liquid into the atmosphere (Carey, 2008). Although atmospheric evaporation is very simple, it is very slow and energy inefficient. The chances of product quality deterioration are also very high. Overall, product quality degradation, slow processing, and inefficiency are the disadvantages associated with atmospheric evaporation. The main objective of evaporation, i.e., application of heat to remove water or volatile components from the main product, needs to be carried out in a technically appropriate way. A key criterion is raising the temperature to the boiling point of the liquid or solvent that is going to be separated from the main product. However, care must be taken to protect the product from thermal degradation during evaporation (Kerr, 2013). Decreasing the boiling point of the water or solvent by decreasing the pressure is one possible way of protecting the product from thermal degradation, ensuring the retention of flavors and nutrients that are sensitive to high temperature (Kerr, 2013). Apart from pressure, the boiling point of liquid food solvents also depends on the concentration of the solute. Normally, the presence of solute increases the boiling point of pure water. Solute concentration is therefore a key parameter for deciding the operating pressure for any liquid during evaporation. The combination of operating pressure and solute concentration will help to decide the boiling point of a particular solution. The Duhring chart is widely used to determine the boiling point of a solution at various operating pressures (Kerr, 2013). The boiling temperature of water at specified pressure can be determined from a steam table, and the boiling temperature of the corresponding solution at that pressure can easily be determined from the chart.

Evaporation plants are in widespread use across the world for the concentration of liquids. Various types of evaporator have been developed with the main aim of maintaining the quality of the final product and to avoid any kind of damage. Common to all evaporators is the use of steam as the heating medium. This chapter describes some of the most widely used evaporators, with their working principles and applications.

3.2 PRINCIPLES AND CLASSIFICATION OF EVAPORATORS

Energy and mass balance are the two basic principles involved in evaporation, along with the estimation of the boiling point elevation of the concentrated liquid. The process largely depends on the heat required to evaporate water or a volatile liquid, the maximum permissible temperature of the liquid, the optimum pressure for evaporation, and any changes that might occur in the main product during evaporation.

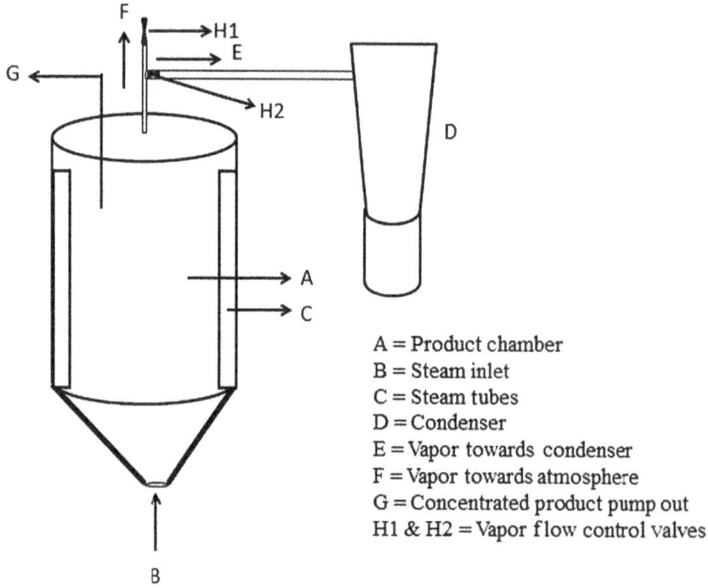

FIGURE 3.1 Pan and batch evaporator

3.2.1 Pan and Batch Evaporators

The pan and batch evaporator is one of the simplest types of evaporator (Figure 3.1). A chamber or pan containing the food item is surrounded by a number of tubes or plates which act as a heat exchanger and carry hot steam. While evaporation is occurring, vapor is released either directly into the environment or directed to the condenser. Vapor flow control valves are attached for this purpose. After the product concentration has reached the desired level, heating is stopped and the product is pumped out of the chamber. Evaporation mostly takes place at ambient pressure, except for heat-sensitive products, for which the evaporator is equipped with a vacuum system. A condenser is fitted, usually in a higher position, to provide the vacuum system. This technique removes common pathogenic microbes.

3.2.2 Short Tube Evaporators

The minimal rate of heat transfer due to the small heat exchange area is a serious drawback of pan and batch evaporators because the product has to remain so long inside the product chamber. This problem can be overcome by using a short tube evaporator (Figure 3.2). This type of evaporator consists of a vertical pipe, with a series of tubes or pipes installed vertically inside the feeding chamber between the liquid and the heating medium to improve the rate of heat transfer. Short tube evaporators are low-cost and simple in construction, and cleaning and maintenance are easy. Inside the chamber feed falls through the passage between the tubes and the final concentrated product is collected from the bottom. Vapor escapes through the upper outlet and is directed to the condenser for condensation. The steam tubes are

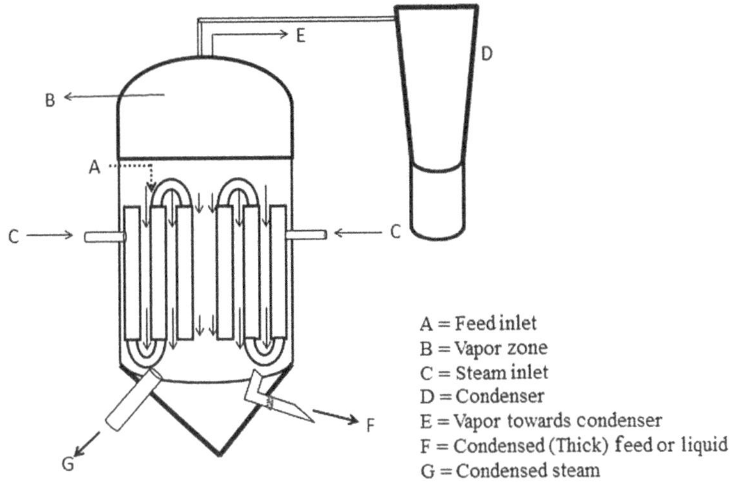

FIGURE 3.2 Short tube evaporator

normally 1.2 m long; length-to-diameter ratio varies from 20 to 40 (Coulson et al., 1991). The time spent in short tube evaporators is comparatively long, so this type of evaporator is commonly used for concentration processes such as sugar cane juice concentration or in salt manufacturing (Kerr, 2013).

3.2.3 Rising-Film Evaporators

The rising-film evaporator (Figure 3.3) is the most commonly used vertical pipe evaporator. A series of long vertical pipes or tubes is installed inside an enclosed

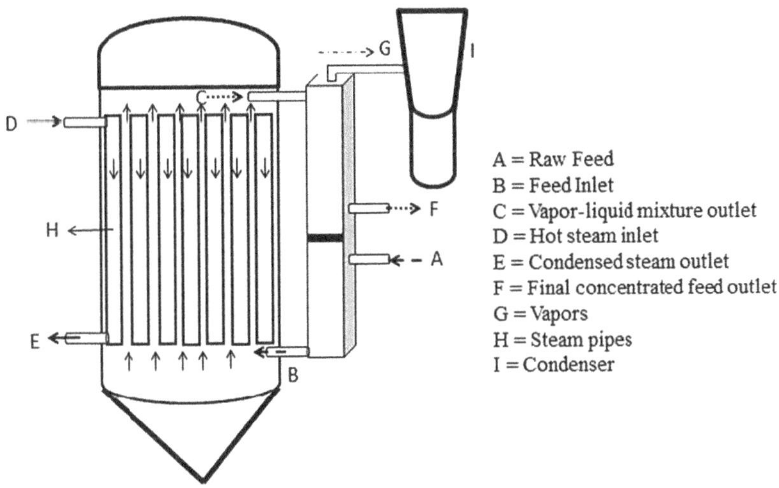

FIGURE 3.3 Rising-film evaporator

chamber to carry the steam as heating medium from top to bottom, and the feed enters from the bottom. The tubes are 2–5 cm in diameter and 10–15 m long (Kerr, 2013). At the lower end of the chamber the liquid feed enters a high-temperature zone, causing bubbles to start forming along its journey from bottom to top. This activity makes a film of liquid ascend the steam pipes, warming the feed rapidly. At the top of the chamber concentrated liquid is separated from vapor, due to the weight difference between the liquid and the vapor. Steam exiting the steam pipes at the top proceeds to the condenser at the bottom of the main chamber. The residence time in this type of evaporator is much shorter, and the floor space requirement is also less, although it requires significant height space. A temperature difference of at least 15°C is required between the feed and steam to generate adequate rising activity. This type of evaporator is appropriate for low-viscosity liquid, as the fluid film needs to move counter to the gravitational force.

3.2.4 Falling-Film Evaporators

The feed movement in the falling-film evaporator (Figure 3.4) is the opposite of that in the rising-film evaporator. Steam movement, however, is the same as in the rising-film evaporator. Due to gravity, the rate of liquid flow in the falling-film evaporator is much faster than in the rising-film evaporator. The liquid–vapor separator, where the vapor is separated and directed to the condenser, is attached at the bottom. The vapor separator is usually associated with a vacuum system for simple partition of vapors. With the extra benefit of gravitational flow, falling-film evaporators are more suitable for viscous fluids. But due to the low residence time (20–40 s), recirculation of the liquid sometimes becomes necessary for proper concentration. The falling-film evaporator is also appropriate for heat-sensitive products like milk and fruit juices.

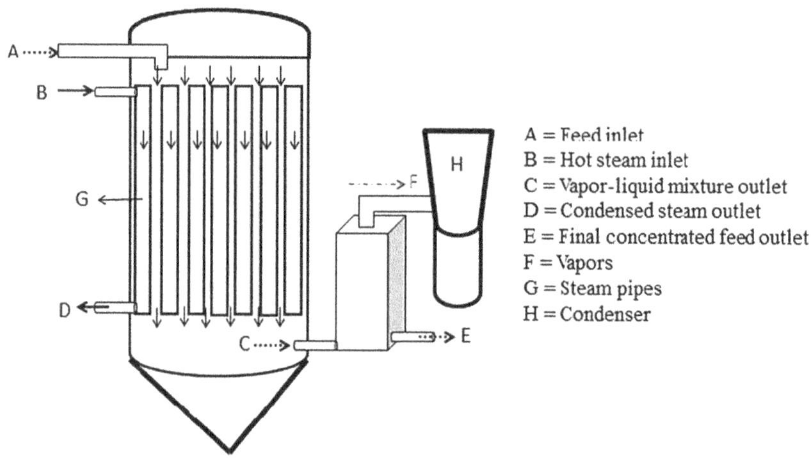

FIGURE 3.4 Falling-film evaporator

Design of this type of evaporator is complex since the hot liquid needs to move downwards. At the same time the tube walls should be covered with fluid film to avoid the risk of a low heat transfer rate. In addition, a skilled and experienced operator is required, because it is very difficult to change the viscosity of the product, the rate of feed flow and the operating temperature while the falling-tube evaporator is in operation.

3.2.5 Agitated Film Evaporators

Creamy vegetables and fruit pastes that are highly viscous and solid in nature have a tendency to become fouled in the high heating zone and get stuck on the wall surface. Fouling is the development of a layer of deposits over time on the heat transfer surface, causing the evaporation and heat transfer rates to reduce, which may affect the quality of the final concentrate (Minton, 1986). Rising- and falling-film evaporators are therefore not appropriate for these kinds of products. Continuous stirring is required to spread these products as a film on the wall and to keep the wall surface clean by regular sweeping. The agitated film evaporator (Figure 3.5) is one of the best options for this purpose. In the agitated film evaporator feed enters the evaporation chamber from the top, and a rotor with multiple blades installed inside the chamber continuously mixes the liquid and prevents fouling of the chamber wall surface. Continuous stirring by the rotor also maintains the heat transfer rate. Movement of steam and feed in the agitated film evaporator is similar to that of the falling-film evaporator. The residence time is normally less than 2 min (Kerr, 2013), and both concentrated product and condensed steam exit the chamber thorough the bottom section.

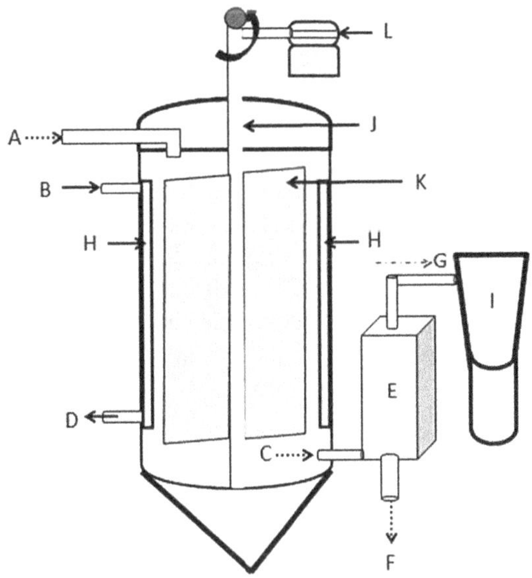

A = Feed inlet
B = Hot steam inlet
C = Vapor-liquid mixture outlet
D = Condensed steam outlet
E = Vacuum chamber
F = Final concentrated feed outlet
G = Vapors
H = Steam jacket
I = Condenser
J = Rotor
K = Scrapers
L = Motor

FIGURE 3.5 Agitated film evaporator

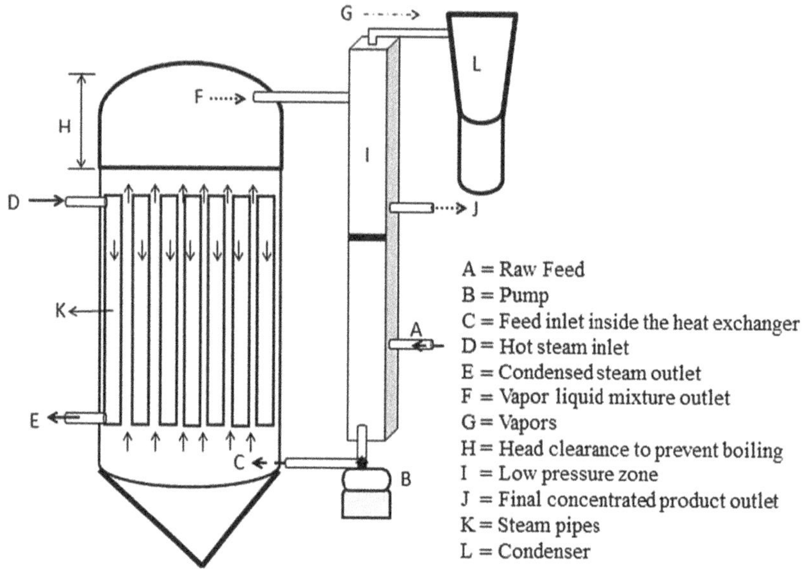

FIGURE 3.6 Forced circulation evaporator

3.2.6 Forced Circulation Evaporators

The forced circulation evaporator (Figure 3.6) is another option for reducing the fouling effect of highly viscous liquids. However, both operation and maintenance costs are high. In the forced circulation evaporator, also known as a calandria evaporator, an external pump is used to recycle the feed through the external heat exchanger (Rao, 2016). The feed is commonly heated a couple of degree for each pass through the external evaporator, so that with high flow velocity (normally 2–5 m/s), a high-capacity pump is required. The concentrated liquid is separated from vapor in the vapor separator placed over the heat exchanger where concentration has taken place. The concentrated liquid comes out through the low-pressure zone of the vapor separator, and vapor is directed to the vapor condenser. A certain amount of head clearance is required in the design of the heat exchanger for the forced circulation evaporator to avoid the liquid itself boiling in the tube. The advantage of the forced circulation evaporator is that orientation of the heat exchanger can be vertical or horizontal as per the requirements. Forced circulation evaporators are mainly used for concentrating highly viscous liquid such as tomato paste and fruit puree and for crystallization in the sugar industry.

3.2.7 Plate Evaporators

Plate evaporators are similar to plate heat exchangers with the additional facility to boil viscous liquids and isolate the water vapor created during heating. This type of evaporator follows the same principles as the long tube evaporator, but has the

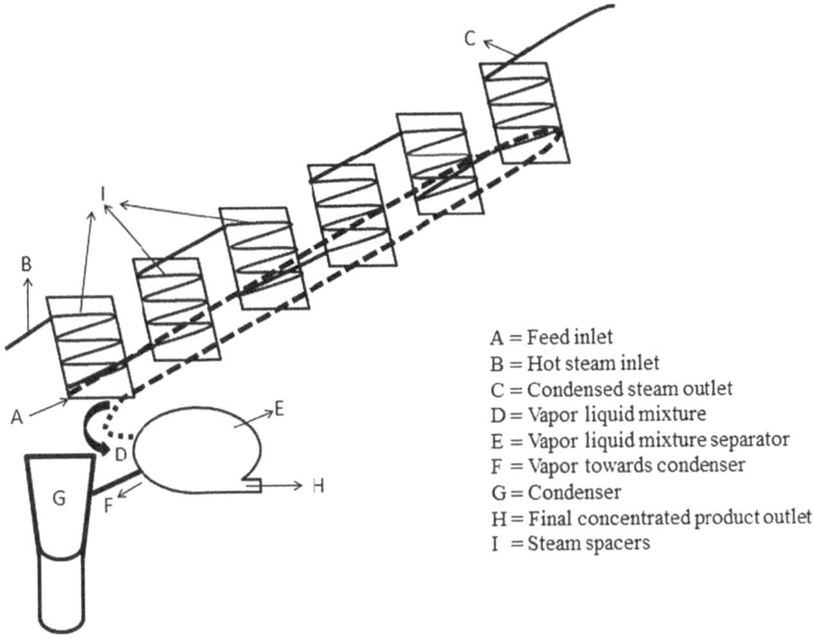

FIGURE 3.7 Plate evaporator

advantage of shorter length, making it easier to install inside the plant building (Earle, 1983). At the same time, it is easy to disassemble, clean, and maintain. Plate evaporators have both steam and product flow arranged in a single plate; a number of plates are connected together with a gasket in between (Figure 3.7). Heating occurs rapidly. After passing through each compartment, the final concentrated liquid arrives in a vapor separator, where it is separated from vapor. The vapor is either directed to the condenser or recirculated for the next cycle. In general, plate evaporators are used to concentrate heat-delicate materials like milk so they are widely used in the dairy industry.

Table 3.1 sets out the merits and demerits that need to be considered in selection of the correct type of evaporator.

3.3 EVAPORATOR OPERATION

3.3.1 Single-Effect Evaporators and Their Application

Single-effect evaporators evaporate vapor from boiling liquid at a single pressure. This is the simplest type of evaporation system and consists of a heating source, a vapor chamber, and a condenser. Liquid feed enters the shell-and-tube type heat exchanger where steam passes through the shell and liquid passes through the tube. Tubes of different sizes and orientations are used according to the requirements. Inside the heat exchanger the operating pressure controls the boiling point of the

TABLE 3.1
Merits and Demerits of Various Evaporators

Type of Evaporator	Merits	Demerits
Pan and batch evaporator	1. Simple design and easy installation. 2. Heat transfer can be increased with the use of agitator.	1. Limited feed at a time. 2. Not applicable for thermal-sensitive food. 3. Low coefficient of heat transfer. 4. High residence time.
Short tube evaporator	1. Utilizes less head space. 2. Suitable for moderately heat-sensitive foods. 3. Good heat transfer coefficient.	1. Not applicable for highly heat-sensitive food. 2. Agitation is required for crystalline food product.
Rising film evaporator	1. Due to the separated phase in partial form gives high heat transfer area. 2. Suitable for foaming liquids.	1. High head space required.
Falling film evaporator	1. Economical to install. 2. Low holding time. 3. Less floor space required.	1. Recirculation limits its use. 2. High headspace required.
Agitated film evaporator	1. Reduced residence time. 2. Rotor gives high coefficient of heat transfer rate. 3. Suitable for viscous foods.	1. High cost of installation. 2. High power usage.
Forced circulation evaporator	1. High coefficient of heat transfer. 2. Low possibility of fouling.	1. Long holding time. 2. More power consumption. 3. Not suited for food which degrades due to long exposure at high temperature.
Plate evaporator	1. Low head space needed. 2. Easy to maintain and suitable for foaming and viscous food.	1. High gasket area required. 2. Chances of leakage.

liquid. Once the liquid temperature reaches boiling point, vapor is separated from liquid in the vapor chamber and the concentrated final product leaves through the bottom of the vapor chamber. As vapor is normally lighter than concentrated liquid, it escapes from the vapor chamber through the vapor side and is either released to the environment or directed to the condenser where it is turned to condensate and returned to the sump for the next run.

Maintaining the mass balance is very important when designing a single-effect evaporator for optimum utilization. Total mass balance is given as:

$$M_{lf} = M_{fp} + M_v \tag{3.1}$$

where

M_{lf} = Mass of liquid feed
M_{fp} = Mass of final product
M_v = Mass of vapor

To achieve the best result there must be balance between the solid content of the mixture entering the heat exchanger and that exiting as a concentrate:

$$X_{lf}.M_{lf} = X_{fp}.M_{fp} \tag{3.2}$$

where
X_{lf} = Solid fraction of liquid feed
X_{fp} = Solid fraction of final product

Together with these balances, the enthalpy balance is also essential to determine the exact quantity of steam required for latent heat of vaporization of water or volatile liquids from feed. A total enthalpy balance is given as:

$$M_{lf}.h_{lf} + M_s.h_s = M_v.h_v + M_{fp}.h_{fp}.M_s.h_{cs} \tag{3.3}$$

where
M_s = Mass of steam
h_{lf} = Enthalpy of liquid feed
h_s = Enthalpy of steam
h_v = Enthalpy of vapor
h_{fp} = Enthalpy of final product
h_{cs} = Enthalpy of condensed steam

The enthalpy of condensed steam only applies if vapor is directed to the condenser rather than released to the environment. It is always preferable to condense the vapor for reuse, saving water and rendering the evaporator more economical in operation.

In addition to design, construction, mass, and energy balance, operational features of the evaporator such as boiling point elevation, initial feed state and viscosity, fouling, corrosion, and foaming also require particular attention. Huge losses can result from changes in liquid characteristics during evaporation (see Section 3.7).

Although single-effect evaporators are simple in construction, require low capital and operating costs, and are easy to operate (Al-Juwayhe et al., 1997), their output is low. They are normally used for small-scale operations or when the expense of steam is generally modest in comparison with evaporator costs. However, for large-scale processes it is always better to use multiple-effect evaporators to reduce steam costs and obtain maximum output.

3.3.2 Multiple-Effect Evaporators and Their Application

In single-effect evaporators, after one-time use steam goes to the condenser for condensation, even though that vapor contains a significant amount of heat. This increases the steam consumption rate and reduces the energy efficiency. Multiple-effect evaporators eliminate this restriction by joining together a series of single-effect evaporators (Figure 3.8) as per the requirement to obtain maximum steam economy (Equation 3.4).

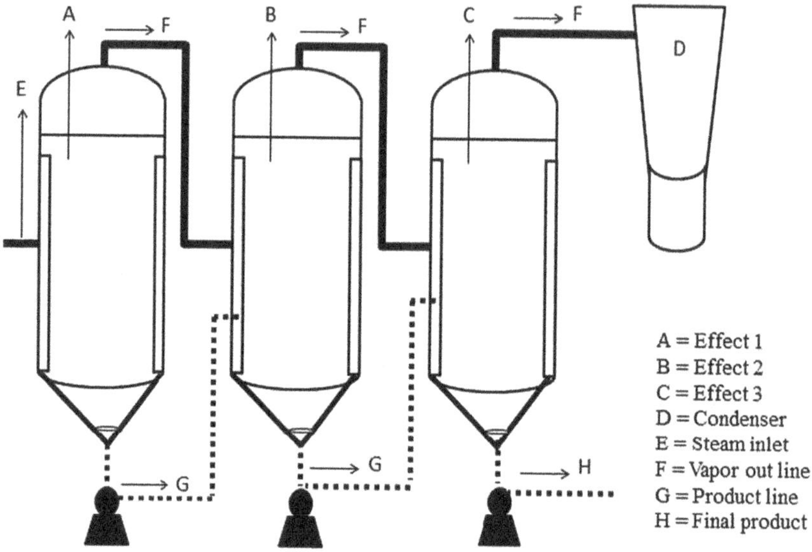

FIGURE 3.8 Multiple-effect evaporation

$$\text{Steam economy} = \frac{\text{Mass of vapor produced}}{\text{Mass of steam consumed}} \tag{3.4}$$

In a multiple-effect evaporator, vapor generated in the first evaporator goes to the second one as a heating medium. In practice, this method is possible if the liquid temperature of the second evaporator is lower than the vapor temperature coming from the first evaporator. The pressure of the second evaporator also needs to be lower than that of the first evaporator to create a positive value of temperature difference over the steam chest surface of the second evaporator. For the rest of the connected evaporators, temperature and pressure should be less than in the previous evaporators, as per Equations (3.5) and (3.6). To maintain this process a vacuum is required in subsequent evaporators to reduce the pressure, which also help the liquid to travel into the subsequent evaporators without the need to install an external pump between them. The heat transfer area of all the evaporators is normally assumed to be equal (Equation 3.7) to provide equal heat transfer rate and economic operation (Equation 3.8). However, it is assumed that all the latent heat of the steam used to heat liquid in evaporator 1 goes to evaporator 2, neglecting the enthalpy required to increase the temperature of the feed.

$$T_1 > T_2 > T_3 > \ldots\ldots\ldots > T_n \tag{3.5}$$

$$P_1 > P_2 > P_3 > \ldots\ldots\ldots > P_n \tag{3.6}$$

$$A_1 = A_2 = A_3 \tag{3.7}$$

$$Q = Q_1 = Q_2 = Q_3 \tag{3.8}$$

Multiple-effect evaporators may have a forward, backward, mixed, or parallel type of feeding mechanism.

In the forward-feeding mechanism, the most dilute item enters the high-pressure stage, and progressively more concentrated items enter the ensuing stages. This is a less complex framework, where product in general flows from first to subsequent evaporators. This type of arrangement is normally used for scale-forming fluids on the grounds that concentrated items are exposed to the lowest temperature. However, they are not suitable for cold feed, as the steam contribution in the first chamber increases the feed temperature and a modest amount of heat will supply the latent heat of vaporization. The lower amount of vapor in evaporator 1 will produce a lower amount of vapor in subsequent evaporators, ultimately reducing the overall economy of the system.

In a backward-feeding mechanism the most dilute feed enters the lowest-pressure evaporator and evaporation takes place at the lowest boiling point. This normally provides a more prominent evaporation limit, as the thickest, hardest to concentrate fluid is presented as most suitable. Because of this, backward-feeding multiple-effect evaporators are suitable for cold and viscous products without affecting the overall economy of their operation. Nonetheless, running costs of backward-feed evaporators are higher than for others, as pumps are required for each evaporator to bring the liquid from the low- to the high-pressure zone.

In a mixed-feed arrangement feed is introduced at an intermediate stage, with liquid initially moving forward and then backward to the first stage and steam introduced in the first. The mixed-feed type is cheaper to run than the backward feed because pumping is only required when the liquid moves from the low- to the high-pressure zone.

In the parallel-feed mechanism the feed is introduced in every stage and steam is introduced in the first stage. This type of arrangement is suitable when feed has to be slightly concentrated.

3.4 ROLE OF EVAPORATION TECHNIQUES IN VARIOUS FOOD-PROCESSING PRACTICES

3.4.1 Concentration of Liquid Foods

Both thermal and non-thermal techniques can be used to concentrate any liquid product, to change its rheological properties such as viscosity, flow behavior, density, and volume, and to slow down processes responsible for deterioration of quality and nutritional value (Bøgh-Sørensen, 2006). In general, increases in the viscosity and density of liquid food are a function of temperature increase, causing phase change and high solid content. Evaporation is the main way of producing concentrate of liquid food. Many considerations are relevant to the concentration of liquid food, including reducing packaging bulk to facilitate handling and transportation, increasing storage life and shelf life by reducing water activity, and increasing product stability (Sorour et al., 2012). Concentration is also used as a pre-feed treatment for processes like drying, crystallization, and mixing.

Evaporation is a widely used thermal technique in concentration that removes water from the raw product which is non-volatile in nature. For heat-sensitive products containing volatile compounds, high temperature can adversely affect product quality. The boiling temperature of heat-sensitive liquid food can be decreased by applying a vacuum which helps concentration without affecting the nutritional balance of the product. During evaporation, water present in liquid food is converted to drops and exposed to hot gas such as hot air which transfers the heat by convection and conduction from the surface to the core of the drop. The moisture is then evaporated from the drop by diffusion and convection to the hot gas. This process is dependent on factors such as evaporation rate, holding time, heat transfer area of the drop, humidity in the evaporating chamber, and relative velocity of the falling drop (Ranz and Marshall, 1952). Concentration is a useful step before freeze drying as it decreases the amount of water needed to freeze, reducing the time and cost of freeze drying. As the concentration of product increases, it has a tendency to fouling which can be overcome by modifying the evaporator design. The evaporation rate is controlled by a combination of time and temperature, in order to obtain the desired concentrate consistency. During this process many desirable volatile compounds are lost due to their volatility, and this is one of the main disadvantages associated with thermal processing. These compounds can be collected using a condenser and collector, and added back to the product.

3.4.2 Crystallization

Crystallization is the process of removal of whole water above the saturation level of solids present in colloidal solution, for example in liquid foods including juice, sugar syrups, milk, and salinated water (Hartel, 2002), principally through evaporation. Spontaneous crystallization occurs when the solute concentration in the solution exceeds the saturation level. This process exploits the property of phase separation of a colloidal solution containing soluble and insoluble components. On evaporation of water, the solute(s) form either crystals or an amorphous lattice. The initial step in this process is nucleation. The formation of crystals starts. Concentrated solute particles aggregate onto the initial crystal nuclei, causing them to grow. The crystallization process depends directly on the flow rate and initial volume, and there is a correlation between the number of crystals formed and the evaporation rate. Factors that can be used to control the rate of evaporation are relative humidity and the saturated region of the reagents (Yu et al., 2012). Crystallizers used for this process are available in batch and continuous types. Crystallization also depends on a number of factors in the liquid food, such as solubility of solute, latent heat of vaporization, latent heat of crystallization, and sensible heat.

Crystallization is commonly used in the food industry for the manufacture of products such as sugar (sucrose), salt, and ice cream. In the sugarcane industry, crystallization is the basic technique used to produce sucrose and other sugars from sugarcane and sugar beet. In salt manufacturing, the conventional production method is crystallization, which is used for liquids that may have been concentrated by evaporators or by natural evaporation, i.e., by sun drying in the field. For protein crystallization, vapor diffusion is a commonly used method that uses solvent evaporation.

3.4.3 Flavor Extraction

Flavors are volatile compounds incorporated in the food matrix that can be separated by evaporating and condensing back using the siphoning principle. Solvent-assisted flavor evaporation (SAFE), which uses solvents such as ethanol, methanol, and acetone, is the most commonly used technique for flavor extraction. It may be used for the extraction of flavors from milk, beer, and food suspension such as fruit pulp. SAFE is used for careful isolation of volatile compounds that have a high boiling point from the food matrix. High recovery is obtained for both high and low boiling-point foods (Miyazawa et al., 2015). SAFE uses a vacuum for distillation of aroma compounds. Hydro-distillation is another technique for flavor extraction. Essential oils are extracted from the food and the extracted concentrated flavor is then added to the food product to retain the flavor.

3.4.4 Water and Solvent Recovery

The evaporator includes a recovery unit which helps to collect back the water from which steam was produced. The main principle used for recovery of water and solvent is condensation. During evaporation, volatile compounds such as aroma and flavor molecules evaporate off and can be recovered by condensation, which is followed by concentration of condensed liquid. Cut-back is an aroma recovery method, in which untreated juice is added to the juice that has had high-temperature treatment, in order to maintain the aroma profile and degrees Brix (MacDowell et al., 1948). Aroma compounds are mostly volatile and evaporate in the initial stage of evaporation. In multiple-effect evaporators, the vapor from the first effect is rich in aroma compounds which are collected and condensed back by cooling with cold water circulation. An additional refrigeration unit is used for the condensation of aroma compounds that are distilled and separated into two phases, aqueous and oil. The oil phase, known as essential oil, is the concentrated form of the aroma, and the aqueous phase, or essence, consists of alcoholic compounds. In milk, volatile carbonyl compounds are the flavor compounds that are recovered under low-temperature and low-pressure conditions for distillation (Wong et al., 1958).

3.4.5 Emulsification

Many compounds have properties that are suitable for the human diet and provide health benefits, such as beta-carotene, fatty acids, natural antioxidants, and numerous other compounds (Moraru et al., 2003). These compounds are insoluble in water only, making it difficult to obtain them in simpler units, so they are made available for consumption and easy digestion in the form of emulsions (Silva et al., 2011). Emulsions are prepared in two forms – water – in – oil and oil – in-water – using a high-energy emulsification-evaporation technique in two stages: emulsification of the compound followed by evaporation of emulsion to remove organic solvent. In the first stage, water is added to the compound to form an oil-in-water emulsion. Stabilizers are also added for the stability of the emulsion and to increase shelf life.

The emulsion thus prepared then passes through a rotary evaporator for removal of the organic solvent. As the solvent evaporates, the desired material forms droplets in a particular range of diameters, e.g., for nano-emulsions the diameter is measured in nm. Centrifugation is then performed to collect these particles, followed by washing to remove excess material such as free stabilizer or untrapped compounds.

3.5 INDUSTRIAL APPLICATIONS OF EVAPORATORS

3.5.1 Dairy Industry

Evaporators are used in the processing of products like skimmed milk, condensed milk, yogurt, and milk powder. The most commonly used evaporators are of the circulation, multiple-effect, falling-film, and horizontal tube types (Goff, 2019). Circulation evaporators are used where moderate concentration is required, for example, in yogurt production milk is concentrated before being further processed. This also helps to remove undesirable odors from milk. The extent of concentration is decided by the viscosity and solute concentration requirements for further processing. In skimmed milk powder production, concentrated skimmed milk is used as a feed for drying.

3.5.2 Fruit and Vegetable Industry

Fruits and vegetables are seasonal, varying from region to region. The moisture content of fruits and vegetables ranges from 80 to 90% (Jayaraman and Das Gupta, 1992), so long storage life is difficult to achieve due to high levels of water activity. Processing and product formulation are required for storage and to increase shelf life. Examples include fruit juice concentrate, jam, jellies, tomato puree and paste, dried vegetables, dried fruits, and vegetables juices. Common evaporators used are the unstirred or open-pan evaporator, agitated thin-film evaporator, falling-film long evaporator, and plate evaporator. Pan or kettle evaporators are generally used for the processing of sauces and jam, while falling-film evaporators are used for the processing of fruit juices.

3.5.3 Waste Water Treatment

Vacuum evaporators are used for the treatment of waste water, which is processed and divided into two streams: high-quality water and solid waste. The waste has low moisture content and water obtained is reusable. This effluent treatment gives up to 90–95% recovery of water, and is energy efficient as it uses vacuum.

3.5.4 Corn Industry

Evaporators are used for the deep processing of corn. Multiple-effect, falling-film, and forced circulation evaporators are commonly used. These are applicable for high-viscosity and concentrated foods. Corn steep water is evaporated to the required concentration after the steeping of corn.

3.6 INDUSTRIAL CHALLENGES AND SCALE-UP OF EVAPORATORS

During the separation phase of evaporation, liquid food present in the solvent starts precipitating or concentrating, causing solute particles to stick to the wall of the evaporation chamber or pipelines, which is known as fouling. Fouling is one of the biggest problems during evaporation, as it reduces the heat transfer surface area and, consequently, rate. The evaporator temperature therefore needs to be scaled up to maintain the internal temperature of the evaporation chamber, ultimately burning fouled material if the evaporator is in continuous operation. Due to fouling, more frequent washing than normal is required, which adds to the processing time. Since fouling is also dependent on the construction material, the evaporator should be made of a material that solute does not stick to and is not corrosive. Most liquid foods are acidic or can generate acidic components during high-temperature reactions. Acid that is present may react to the evaporator material, degrading the surface as well as causing harmful changes in the product such as the addition of metal, or color degradation compounds (Jellesen et al., 2006). Solutions include using food-grade nonreactive metal coating, and cleaning with non-caustic cleaners which clean the surface efficiently but do not contribute to surface pitting or corrosion. Selection of cleaning agents is also a considerable factor in maintaining the life of evaporators. In the food industry, stainless steel is mostly used as it is safe for food purposes (Jellesen et al., 2006).

Evaporators work in two modes: batch or continuous. Batch evaporators are separate units in which one lot at a time is evaporated, while in continuous mode the feed continuously runs through the evaporator. Continuous evaporators are efficient and economical to run, as they utilize the same steam for successive evaporation chambers. In the batch process, the steam is utilized only once, so each new lot wastes power used for steam production. Multiple-effect evaporators that use the same steam for every step also use low pressure, which maintains the temperature by reducing the boiling point of the feed due to the pressure drop, although this can adversely affect product nutritional value due to the high-temperature processing and long retention time. These losses, in turn, can be overcome by recovery and by reducing temperature, glass transition temperature, retention time, and boiling point. Modern evaporators have considerable advantages over the traditional type as regards heat transfer area, reduced holding time, reduced space requirement, continuous feed, use of scrappers, and use of baffles to reduce fouling. Evaporators, being both traditional and cheaper, are more widely used in industry than other equipment using thermal or non-thermal techniques.

Changes in product quality caused by high temperature have led to technical advances in the design and operation of evaporators. Quality degradation, such as inversion of sugar, protein denaturation, loss of vitamins and flavor, can be overcome through technical advances such as low temperature, low holding time, and the use of low pressure or vacuum for heat-sensitive products.

Table 3.2 shows industrial problems related to evaporators, together with their possible reasons and preventive measures applied to scale up efficiency, performance, and productivity.

TABLE 3.2
Industrial Problems of Evaporators, Their Possible Reasons and Prevention

Problems	Reason	Prevention
Fouling on evaporator wall	1. Caused by chemical reaction within the fluid, growth and deposition of organisms on evaporator wall surface. 2. Mainly due to high temperature.	1. Controlled by controlling the formation of activated serum protein molecules. 2. Installation of holding tank with high volume-to-surface ratio (de Jong et al., 1992). 3. De-gassing of liquid feed.
Glass transition temperature (T_g)	Sensitive compounds have very low glass transition temperature.	Use of carrier agents to increase the T_g of liquid food, such as maltodextrin, xanthum gum (Shrestha et al., 2007)
High head space required	Installation of vertical long tube rising-film or falling-film evaporators for low-to-moderate viscosity products such as milk and fruit juices.	Building construction with high head space and minimum floor space.
Loss of nutrients, aroma, and flavor	Oxidation of lipids, volatility of aroma compounds, solubility of vitamins and darkening of color or loss of color due to high temperature and retention time.	Losses can be reduced by reducing time and temperature combination.
Sugar inversion	1. High residence time and high temperature and larger surface area (Eggleston et al., 2004). 2. Acid catalyzed inversion due to drop in pH because of dead volume of evaporator.	1. Can be reduced by varying temperature and pH according to the process.
Protein denaturation	High temperature and retention time.	1. Changing evaporator design. 2. Reducing the temperature.
Heat and energy loss	Batch process of evaporation.	Using multiple-effect evaporators.
Corrosion	Acidic nature of the food material.	Using inert food-grade material.

3.6.1 Economic Considerations

The economy of an evaporator is defined as the total quantity of water or volatile liquid that is vaporized per unit mass of input feed. The economy of single-effect evaporators is almost always less than 1, but for multiple-effect evaporators it may be substantially greater (Standiford, 2000). Various factors affect evaporator economy, including heat transfer rate on both the steam and product side, temperature difference between steam and boiling product, initial state of feed, heat transfer coefficients and wall thermal resistance, fouling, corrosion, and foaming. The following sections describe the main factors affecting the economy of an evaporator.

3.6.2 Heat Transfer Rate on Both Steam and Product Side

The rate of heat transfer is one of the key factors responsible for the economic viability of an evaporator. Rate of heat transfer (Q) is the combined effect of three parameters (Equation 3.9): heat transfer area (A); overall heat transfer coefficient (U); and overall temperature difference (ΔT).

$$Q = UA(\Delta T) \tag{3.9}$$

A drop in any of these parameters ultimately affects the heat transfer rate, resulting in reduced economy of the evaporator.

3.6.3 Temperature Difference between Steam and Boiling Product

If the solutes are small, insoluble colloidal material, the boiling temperature of an aqueous solution in the vapor space of an evaporator at a given pressure will be equal to the boiling temperature of pure water. But if the solute is soluble in aqueous solution, boiling temperature will be more than pure water, which is called boiling point elevation. Boiling point elevation reduces the usual temperature difference between the heating medium and the liquid, which ultimately reduces the capacity of the evaporator as well as affecting its economic operation. Liquid head is another important factor affecting evaporator economy. The measure of decrease cannot be exactly evaluated quantitatively; however, the subjective impact of fluid head, particularly with high liquid levels and high fluid speed, ought not to be overlooked.

3.6.4 Initial State of Feed

Initial heat conditions of the feed also determine the economic status of an evaporator. During entry to the heat exchanger, the feed temperature should not be below or above boiling point. If feed is in a cold state, extra heat will be required to bring its temperature up to boiling point, reducing evaporator capacity and causing it to operate less economically. On the other hand, if the feed temperature is above boiling point, a proportion of the feed will be evaporated in what is known as 'flash evaporation', also resulting in economic loss. So, to achieve appropriate performance from an evaporator, the feed must enter at the boiling point corresponding to the absolute pressure in the vapor space.

3.6.5 Heat Transfer Coefficient and Wall Thermal Resistance

Heat transfer coefficient and wall thermal resistance strongly influence the economic sustainability of an evaporator. Resistance to heat transfer from heating steam to liquid is affected by steam film resistance, and the presence of non-condensable gas reduces the steam film coefficient. Tube wall resistance and resistance from boiling liquid also reduce the heat transfer coefficient, leading to reduced evaporator performance. Design, material selection, and operational conditions are the parameters that need to be controlled if optimum heat transfer coefficient is to be achieved for improved evaporator performance.

3.6.6 Fouling and Corrosion

Fouling reduces the overall heat transfer coefficient and causes corrosion which may adversely affect the thermal and mechanical performance of heat exchangers in an evaporator, leading to economic losses (Su et al., 2015). Design considerations may help to reduce the losses due to fouling. These include application of more fouling fluid to the tube side, coating an artificial fouling layer, maintaining constant fluid velocity, simple access for cleaning, and maintaining the tube wall temperature to prevent salt deposition during water flow.

3.6.7 Foaming

Foaming is a common problem associated with the evaporation process. Foaming occurs due to the application of high amounts of heat, vacuum or rotational speed, leading to product loss. Foaming cannot be totally eliminated but its effects can be minimized by expert handling and observation, and proper control of temperature, vacuum, or rotational speed. Opening the upper stop cock whenever foam appears is one of the common techniques to control foam during evaporation.

3.7 RESEARCH-ORIENTED PROBLEMS

Controlling the final concentration of the product leaving the evaporator is vital, as this defines the overall performance of the evaporator. In order to obtain a final product of consistent quality, problems that arise during processing must be resolved before they have a chance to affect the whole of the solid concentrate (Bakker et al., 2006). Complexities arise during evaporator operation due to their physical structure, sequential arrangement, and procedure adopted for their function (Karimi et al., 2007; Rousset et al., 1989). Changes in operating parameters are a huge research challenge, which needs to be addressed without changing standard concentration of solid output. Operational parameters such as evaporator size, heat transfer rate, temperature, and residence time are the major challenges as they affect both energy consumption and final overall product quality. Engineers and researchers are seeking solutions with the development of various modifications, mathematical models, and new instruments (Kaya and Sarac, 2007) with the aim of achieving optimum residence time, temperature, and heat transfer rate, as well as maintaining economic performance.

Research has shown that pressure cleaning and capturing spillages in multiple-effect evaporators can reduce energy consumption by 40% (Salakki et al., 2014). Outlet temperature can be increased by increasing the steam tube pitch (Parikh and Patel, 2013) and then decreased under given working conditions. It has also been found that the heat transfer coefficient increases with increasing pressure (Yu et al., 2012), although there is a lack of data on water boiling in an evaporator at sub-atmospheric pressure. Karimi et al. (2007) developed an inferential cascade control for use in a multi-effect falling-film evaporator to improve control performance, decrease time response in the control loop, and to estimate precisely process states that cannot be measured. Miranda and Simpson (2004) developed a strong model that can give precise process variable values in spite of the parametric vulnerability.

Further research and development is required in the areas of equipment maintenance and preservation, control of parameters, and optimization of effects for improved efficiency and economy.

3.8 CONCLUSION

Evaporation is a mature technology that produces excellent processed food by removing water or volatile substances from food products. Evaporators may be of single-stage or multiple-stage types. As the number of stages increases, evaporator operation also becomes more economical. The optimal number of stages needs to be determined to establish an economic balance between energy saving and additional investment. Evaporation provides food microbiological stability, increases the shelf life of the product, and reduces transportation and storage costs. Today's discerning consumers, however, are very concerned about the nutritional quality and freshness of food, which presents challenges for evaporators. Research is ongoing to develop and modify evaporators to produce minimally processed food with the least change in the final product. Concern over energy and consumption costs is giving rise to a continuing search for improved procedures and structures to limit the huge energy requirements of evaporation.

REFERENCES

Al-Juwayhel, F., H. El-Dessouky, and H. Ettouney. 1997. Analysis of single-effect evaporator desalination systems combined with vapor compression heat pumps. *Desalination* 114(3): 253–275.

Al-Muhtaseb, A.H.W., A.M. McMinn, and T.R.A. Magee. 2002. Moisture sorption isotherm characteristics of food products: A review. *Food Bioprod. Process* 80(2): 118–128.

Bakker, H.H.C., C. Marsh, S. Paramalingam, and D. Chen. 2006. Cascade controller design for concentration control in a falling-film evaporator. *Food Control* 17: 325–330.

Bøgh-Sørensen, L. 2006. *Recommendations for the Processing and Handling of Frozen Foods*. Paris: International Institute of Refrigeration (IIR).

Carey, V.P. 2008. *Liquid-Vapor Phase-Change Phenomena: An Introduction to the Thermophysics of Vaporization and Condensation Processes in Heat Transfer Equipment*. London: Taylor and Francis.

Coulson, J.M., J.F. Richardson, J.R. Backhurst, and J.H. Harker. 1991. *Chemical Engineering*. Oxford: Pergamon Press.

de Jong, P., S. Bouman, and H.J.L.J. van der Linden. 1992. Fouling of heat treatment equipment in relation to denaturation of fl-lactoglobulin. *J. Soc. Dairy. Technol.* 45(1): 3–8.

Dechow, J.F. 1989. *Separation and Purification Techniques in Biotechnology*. New Jersey: Noyes Publications.

Earle, R.L. 1983. *Unit Operations in Food Processing*. Toronto: Pergamon Press.

Eggleston, G., M. Damms, A. Monge, and T. Endres. 2004. New insights on sucrose losses across factory evaporators and juice and syrup clarifiers. *Proc. Sug. Proc. Res. Conf.* 349–370.

Goff, H.D. 2019. Dairy product processing equipment. In: K. Myer (ed.), *Handbook of Farm, Dairy and Food Machinery Engineering*. New York: Elsevier.

Hartel, R.W. 2002. Crystallization in foods. In: A.S. Myerson (ed.), *Handbook of Industrial Crystallization*. New York: Butterworth-Heinemann; Elsevier, pp. 287–304.

Hayashi, H. 1989. Drying technologies of foods – Their history and future. *Dry. Technol.* 7(2): 315–369.

Jayaraman, K.S. and D.K. Das Gupta. 1992. Dehydration of fruits and vegetables – Recent developments in principles and techniques. *Dry. Technol.* 10(1): 1–50.
Jellesen, M.S., A.A. Rasmussen, and L.R. Hilbert. 2006. A review of metal release in the food industry. *Mater. Corros.* 57(5): 387–393.
Karimi, M., A. Jahanmiri, and M. Azarmi. 2007. Inferential cascade control of multi-effect falling-film evaporator. *Food Control.* 18: 1036–1042.
Kaya, D. and H.I. Sarac. 2007. Mathematical modeling of multiple-effect evaporators and energy economy. *Energy* 32: 1536–1542.
Kerr, W.L. 2013. Food drying and evaporation processing operations. In: M. Kutz (ed.), *Handbook of Farm, Dairy and Food Machinery Engineering*. London: Elsevier, pp. 317–354.
MacDowell, L.G., E.L. Moore, and C.D. Atkins. 1948. Method of preparing full-flavoured fruit juice concentrates. U.S. Patent 2, 453, 109.
Minton, P.E. 1986. *Handbook of Evaporation Technology*. New Jersey: Noyes Publications.
Miranda, V. and R. Simpson. 2004. Modelling and simulation of an industrial multiple effect evaporator: Tomato concentrate. *J. Food Eng.* 66: 203–210.
Miyazawa, M., Y. Nakashima, H. Nakahashi, N. Hara, H. Nakagawa, A. Usami, and W. Chavasiri. 2015. Volatile compounds with characteristics odor of essential oil from magnolia obovata leaves by hydrodistilation and solvent-assisted flavor evaporation. *J. Oleo Sci.* 64(9): 999–1007.
Moraru, C.I., C.P. Panchapakesan, Q. Huang, P. Takjistov, S. Liu, and J.I. Kokini. 2003. Nanotechnology: A new frontier in food science. *Food Technol.* 57 (12): 24– 29.
Parikh, K. and T. Patel. 2013. Analysis and validation of fin tube evaporator. *Int. J. App. Innov. Eng. Manag.* 2(3): 430–443.
Ranz, W.E. and W.R. Marshall. 1952. Evaporation from drops. *Chem. Eng. Prog.* 48(3): 141–146.
Rao, D.G. 2016. *Fundamentals of Food Engineering*. New Delhi: PHI Learning Private Ltd.
Rousset, F., Y. Saintcir, and M. Daclin. 1989. Automatic process control of multiple effect evaporation part II: Practical realization and results. *Sugar Ind.* 114(6): 470–476.
Salakki, S., M.A. Lourdu, R. Antony, J.H. Patil, and V. Shetty. 2014. Improving the efficiency of multiple effect evaporator to treat effluent from a pharmaceutical industry. *Int. J. Innov. Res. Sci. Eng. Technol.* 3(7): 14727–14733.
Seader, J. and M. Henley. 1998. *Separation Process Principles*. New York: Wiley.
Shrestha, A.K., T. Ua-arak, B.R. Adhikari, T. Howes, and B.R. Bhandari. 2007. Glass transition behavior of spray-dried orange juice powder measures by differential scanning calorimetry (DSC) and thermal mechanical compression test (TMCT). *Int. J. Food Prop.* 10: 661–673.
Silva, H., M. Cerqueira, B. Souza, C. Ribeiro, M. Avides, M. Quintas, J. Coimbra, M. Carneiro-da-Cunha, and A. Vicente. 2011. Nanoemulsions of [beta]-carotene using a high-energy emulsification-evaporation technique. *J. Food Eng.* 102(2): 130–135.
Somogyi, L.P. and B.S. Luh. 1986. Dehydration of fruits. In: J.G. Woodroof and B.S. Luh (eds.), *Commercial Fruit Processing*. Netherlands: Springer, pp. 353–354.
Sorour, M.A., S.R. Mostafa, and S.M. Bo Samri. 2012. A case study of orange juice concentration by recompression evaporation techniques. *J. Food Process Eng.* 36(3): 337–342.
Standiford, F.C. 2000. Evaporation. In *Kirk-Othmer Encyclopedia of Chemical Technology*. John Wiley and Sons Inc.
Su, J., M. Ma, T. Wang, X. Guo, L. Hou, and Z. Wang. 2015. Fouling corrosion in aluminum heat exchangers. *Chinese J. Aeronaut.* 28(3): 954–960.
Sunderland, T.C.H. 2011. Food security: Why is biodiversity important? *Int. Forest. Rev.* 13(3): 265–274.
Wong, N., P.S. Patton, and D.A. Forss. 1958. Methyl ketones in evaporated milk. *J. Dairy Sci.* 41: 1699–1705.
Yu, Y., X. Wang, D. Oberthur, A. Meyer, M. Perbandt, L. Duan, and Q. Kang. 2012. Design and application of a microfluidic device for protein crystallization using an evaporation-based crystallization technique. *J. Appl. Crystallogr.* 45: 53–60.

4 Blanching Treatments in Food Processing

S.B. Kalse
Department of Post-Harvest Engineering, Post Graduate Institute of Post Harvest Management Dr. Balasaheb Sawant Konkan Krishi Vidyapeeth Dapoli, Maharashtra, India

C.B. Khobragade
Department of Agricultural Engineering, Triguna Sen School of Technology, Assam University, Silchar, Assam, India

CONTENTS

4.1 INTRODUCTION

Blanching is a preliminary heat treatment that inactivates senzymatic activities in fruit and vegetables prior to freezing, drying, or canning. It not only preserves food quality but also expels trapped air, reducing the volume of the material and its microbial load, and eliminating unwanted odors and flavors (Binsi et al., 2014). Blanching is done at temperatures as low as 50°C for up to 40 min (Reis et al., 2008;

Fernández et al., 2006). Incomplete blanching causes critical damage to food because the heat is only sufficient to disrupt tissues and release enzymes but not sufficient to mix the enzymes and substrates, inactivating enzymes that cause deterioration.

Blanching is performed most often by hot water, with steam treatment the next most frequently performed. The blanched product is cooled in water to inhibit thermophiles and avoid a cooked taste. Leaching losses are greater in hot-water blanching than in steam blanching. The blanching process increases the drying rate of many fruits, e.g., peach slices (*Prunuspersica* L.), red pepper (*Capiscum annum*), and carrot cubes (*Daucuscarota* L.) (Kingsly et al., 2007; Akanbi et al., 2003; Turhan and Sahbaz, 1997; Mazza, 1983). To avoid nutritional losses during blanching it is necessary to use newer technologies such as microwave blanching, ohmic blanching, and infrared blanching. This chapter describes novel blanching methods and their effect on the nutritional and textural quality of food material.

4.2 TRADITIONAL BLANCHING METHODS

4.2.1 Hot-Water Blanching

In this method food is treated by hot water at a temperature of 70–100°C for a specific time period, then further dewatered and cooled by water. In industrial blanching, the product is either transported to the hot-water tank by a screw or a chain conveyor, or immersed in a rotary drum. The cylindrical chamber of the hot-water blancher consists of a heating jacket and a layer of insulation; the large diameter screw is immersed in half of the vessel. The water is heated by steam through a heat exchanger. The longer the blanching time, the more leaching of nutrients and minerals from the product occurs (Dauthy, 1995).

4.2.2 Steam Blanching

The product is carried on a mesh conveyor through a steam atmosphere in a tunnel 15 m long and 1.5 m wide where it is injected with food-grade steam (100°C). The residence time of the food is controlled by the speed of the conveyor and the length of the tunnel. Steam losses can be reduced by using sprays at the inlet and outlet to condense escaping steam, and using rotary valves or hydrostatic seals for entry and exit of the food. The heat transfer coefficient of steam is greater than that of hot water, so steam blanching requires less time than water blanching and fewer nutrients are lost. Table 4.1 shows the advantages and limitations of steam and hot-water blanching systems.

4.3 NEED FOR BLANCHING

Blanching treatments are used to improve taste by removing the bitterness of fruits and vegetables and also to improve the color of vegetables. Surface dirt and bacteria can be removed from vegetables by blanching, so it helps to retain their freshness. The different purposes of blanching are shown in Figure 4.1.

TABLE 4.1
Advantages and Limitations of Steam and Hot-Water Blanchers

Method	Advantages	Limitations
Steam blanching	• Less leaching and more higher product yield • Less waste generated and lower disposal costs • Energy efficient • Improved product appearance retaining color, flavor, and texture	• Washers required for adequate cleaning • More food in tunnel leads to uneven blanching • Some loss of mass in the food • Higher maintenance cost • Cleaning is difficult
Hot-water blanching	• Requires less capital investment • Uniform heating of product • Less floor area utilized	• More waste water generated, increasing waste management cost • More susceptible to attack by thermophilic bacteria • More physical damage to products due to turbulence

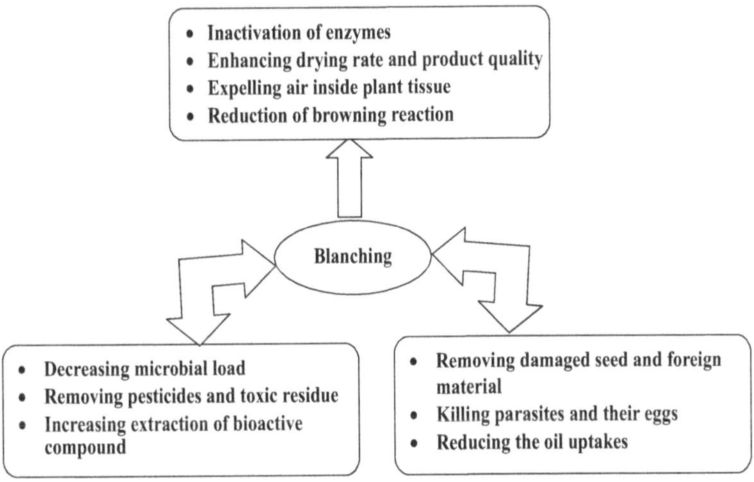

FIGURE 4.1 Purposes of blanching

The blanching process varies slightly depending on the type of fruit or vegetables (see Table 4.2).

4.3.1 Inactivation of Enzymes Related to Food Quality

The deterioration of fruit and vegetables during storage and transportation depends to a great extent on enzymatic reaction (Dorantes-Alvarez et al., 2011). Enzymatic browning causes discoloration of fruit by increasing the concentration of polymeric derivatives of o-quinones, through oxidative reactions catalyzed by polyphenol

TABLE 4.2
Process Conditions, Treatment Details, and Heating Times for Different Blanching Techniques

Blanching Type	Conditions	Treatment	Heating Time	References
Infrared blanching	Medium IR (peak): 3–4.7 μm. Far IR (peak): 6 μm	3000–4000 W/m^2	20 min	Zhu and Pan (2009)
Microwave blanching	Frequency 2450–50 MHz	Temperature 85–87°C	10–11 min	Boyes et al. (1997)
High-pressure blanching	Pressure 100 and 200 MPa	Temperature 22–24°C (100 MPa) 24–26°C (200 MPa)	10 min (100 MPa) 20 min (200 MPa)	Castro et al. (2008)
Hot-water blanching	Temperature 70, 80, 90°C	–	1 min (70, 80°C) 2.5 min (90°C)	Castro et al. (2008)
Steam blanching	Temperature 100°C	–	15–20 s	Wang et al. (1997)
Hot-gas blanching	Steam usage 1.9–2.9 kg/min	–	14–25 min	United States Environmental Protection Agency (1974)

oxidase (PPO) in the presence of atmospheric oxygen (Del Valle et al., 1998). Blanching inactivates these enzymatic reactions which contribute to undesirable flavors, odors, colors, and textures, as well as the breakdown of nutrients. These enzymes are partially inactivated at temperatures of more than 50°C and an appropriate treatment time, and completely inactivated at temperatures of at least 80°C.

The texture of vegetables depends on the rate of peroxidatic activity on phenolic compounds such as ferulic acid. The inactivation of peroxidase indicates complete destruction of enzymatic activities, therefore loss of peroxidase activity is generally used as an index of proper blanching of vegetables (Del Valle et al., 1998). Table 4.3 shows inactivated enzymes during blanching of different vegetables.

4.3.2 Enhancing Drying Rate and Product Quality

Blanching as a pretreatment accelerates the drying rate for three reasons: (i) surface hardening of the food material decreases so that the rate of water evaporation from the surface increases (Orikasa et al., 2008); (ii) the microstructure of the food is changed due to physical damage, reducing internal resistance to moisture diffusion (Watanabe et al., 2014); and (iii) the internal water permeability is improved due to heat stress which damages the cells (Ando et al., 2016). Fruits like plums and grapes with a natural waxy layer on the surface resist moisture evaporation from the surface during drying. Blanching increases the drying rate by changing the physical properties of the products and reducing energy consumption.

TABLE 4.3
Inactivated Enzymes during Blanching of Different Vegetables

No.	Material	Blanching Technique	Conditions	Inactivating Enzyme	References
1.	Peas, spinach, broccoli, potatoes, and carrots	Hot water and steam	Hot water at 85°C and 87.8°C and with steam at 85° and 96.7°C for up to 3.5 min.	*Listeria monocytogenes* and Salmonella	Ceylan et al. (2017)
2.	Peach	High-pressure processing	Pressures (50–700 MPa) at 25°C (5 and 10 min).	Polyphenoloxidase enzymes	Kingsly et al. (2007)
3.	Mushroom	Microwave blanching	2450 MHz microwave radiation	Polyphenoloxidase (PPO)	Devece et al. (1999)
4.	Broccoli	Microwave blanching	2 min in microwave without water and 6 min with microwave and water	Peroxidase (POD) and lipoxygenase (LOX)	Orak (2006)
5.	Carrot juice	Hot water	5 min in hot water at 95°C	Peroxidase (POD) and catalase	Shivhare et al. (2009)

Blanching has been shown to effectively increase the drying rate of kiwifruit (Yoshida et al., 2014), spinach (Hirota et al., 2000), and cabbage (Nilnakara et al., 2009; Watanabe et al., 2014) and to increase the quality of the final processed product by inhibiting undesirable enzymatic reactions (Nilnakara et al., 2009). The blanching of turmeric in boiling water for 4 min increases the effective moisture diffusivity, affecting the rhizome structure and thus increasing the drying rate (Blasco et al., 2006).

Steam blanching and hot-water blanching both increase the drying rate (Alzamora et al., 2000). Steam blanching is useful to increase the drying rate of basil because it destroys its cell walls (Rocha et al., 1993; Ramesh et al., 1999). It was also observed that there is less resistance to moisture movement during drying and the effective diffusivity of moisture increased by more than two orders of magnitude with steam blanching treatments (Sablani et al., 2011).

4.3.3 Removing Toxic Content and Pesticide Residues

Pesticides like herbicides, fungicides, and insecticides are chemical or biological substances used to kill or remove a specific group of pests and prevent crops from being infested. Traces of pesticides may remain on the surface of the crop for a long time, with potential toxic effects for consumers ranging from mild illness such as headaches and nausea to serious diseases like cancer. Removing pesticide residues before consumption of fruit and vegetables is therefore very important for human health. Blanching is key to removal of these harmful residues.

Blanching, washing, peeling, and cooking reduce harmful residues. Toxins are leached into the washing or blanching water. Peeling removes the outer layer of fruit and vegetables with their higher residue concentration (Bajwa and Sandhu, 2014). The octanol/water partition coefficient (K_{ow}) determines how pesticides transfer into water, and also indicates the ability of pesticides to pass through biological membranes and to accumulate in living tissue (Vighi and Di Guardo, 1995).

Huan et al. (2015) studied the effect of traditional Chinese cooking (washing, blanching, stir-frying, frying, and a combination of cooking methods) on eight different pesticide residues (pyridaben, procymidone, chlorothalonil, difenoconazole, a-cypermethrin, bifenthrin, S-fenvalerate, and k-cyhalothrin) in cowpea. The results showed that washing and blanching could reduce residues with low K_{ow}; stir-frying and frying were more effective in residues with high K_{ow}; and stir-frying and frying concentrated residues with low K_{ow}. Villalobos et al. (2016) showed that dehydrating figs in a convection oven at 40, 60, and 70°C reduced the count of yeasts and mesophilic aerobic bacteria for about 3.0, 2.5, and 2.0 log CFU g^{-1} and 5.0, 4.2, and 3.8 log CFU g^{-1}, respectively; but if blanching treatment was used (100°C for 1 min), the bacteria and yeast counts decreased below detectable level.

4.3.4 Expelling Air Trapped Inside Plant Tissues

Air trapped in plant tissue can be expelled by blanching. This is very important during canning, because blanching reduces the expansion of air during the canning process, as well as reducing the strain on the containers and the risk of misshapen cans and faulty seams. Removing gas leads to improved texture, and softer and more transparent tissue in pear (Chafer et al., 2003). In addition, expelling air from the plant tissue reduces oxidation and the risk of corrosion of the materials used for can manufacturing.

4.4 NOVEL BLANCHING METHODS

Novel blanching technologies are energy efficient, reduce nutrient loss, and are more environmentally friendly. Innovative blanching techniques include high-pressure, microwave, ohmic, and infrared blanching. Their principles, characteristics, current application status, and limitations are identified and discussed in the following sections.

4.4.1 High-Pressure Blanching

High-pressure processing destroys microorganisms and inactivates enzymes without using high temperature and with minimal harmful effects on food quality (Smith, 2011).

The drying kinetics and quality of hot-air-dried potato cubes under high hydrostatic pressure and hot-water blanching was studied by Al-Khuseibi et al. (2005), who concluded that potato cubes treated by high-pressure blanching require less drying time than hot-water blanched potato because high-pressure treatment increases the permeability of the potato cell structure. Alvarez-Jubete et al. (2014) found that high-pressure blanching of white cabbage at 200–600 MPa for 5 min at a temperature of 20 or 40°C was superior in preserving bioactive compounds and antioxidant activity, and resulted in improved retention of color and texture.

4.4.2 Microwave Blanching

Microwaves are electromagnetic waves with wavelengths ranging from 1 to 1 mm that have corresponding frequencies ranging from 300 MHz to 300 GHz (Thostenson and Chou, 1999). In addition to their many applications in radar, radio astronomy, navigation, and medicine, microwaves are also used in food processing.

Microwaves interact with polar water molecules in food, which results in a dielectric heating effect. Microwave heating has several advantages over conventional heating, including high heating rates and short processing times, making it suitable for drying, pasteurization, blanching, thawing, tempering, and baking (Ramesh et al., 1999; Bai et al., 2013; Rastogi, 2012).

Microwave blanching has many advantages: it is a rapid, highly energy-efficient process, easy to install and clean (Lin and Brewer, 2005), it liberates bound antioxidants, and increases antioxidant activity (Hihat et al., 2017). The effect of microwave blanching, steam blanching and hot-water blanching on quality of carrot slices was studied by Kidmose and Martens (1999). In this study microwave blanching was performed with 4 magnetrons (power of 1.25 kW for each magnetron) at 2450 MHz and a conveyor speed of 0.5 m/min; steam blanching was carried out by a steamer at 90°C for 3 min; water blanching was performed in a jacket vessel at 90°C for 4 min. Finally, due to the facilated advantages as stated above, it was also reported that microwave-blanched sample had significantly superior effects from those blanched by steam or hot water.

4.4.3 Ohmic Blanching

In ohmic heating, food material placed between two electrodes acts as an electrical resistance, causing heat to be generated within the food and thus the temperature of the product to increase rapidly (Assiry et al., 2003; Sastry and Barach, 2000). The magnitude of the alternating current passing through the food is dependent on the applied voltage gradient and the electrical conductivity of the material. In this method internal thermal energy can be generated so that it does not depend on heat transfer either through a solid–liquid interface or inside a solid in a two-phase system. The principle of ohmic heating is illustrated in Figure 4.2.

Ohmic heating is useful in various unit operations of the food industry, including blanching, evaporation, dehydration, fermentation, extraction, sterilization, and pasteurization (Farahnaky et al., 2012; Halden et al., 1990). Low heat transfer rates make it difficult to blanch large-volume fruit and vegetables by conventional methods; proper blanching can be achieved by ohmic blanching methods.

The ohmic blanching of mussels (9.15 V/cm, 60 Hz) was compared with conventional blanching (50–90°C) in terms of texture and microbial load (Bastias et al., 2015). It was found that ohmic heating reduced the concentration of lead and cadmium, the cutting strength, and the mesophilic, aerobic and enterobacterial counts. The effect of ohmic heating on microorganisms may be attributed to heat and also to electroporation, i.e., the formation of pores in cell membranes due to the presence of an electric field (Knirsch et al., 2010).

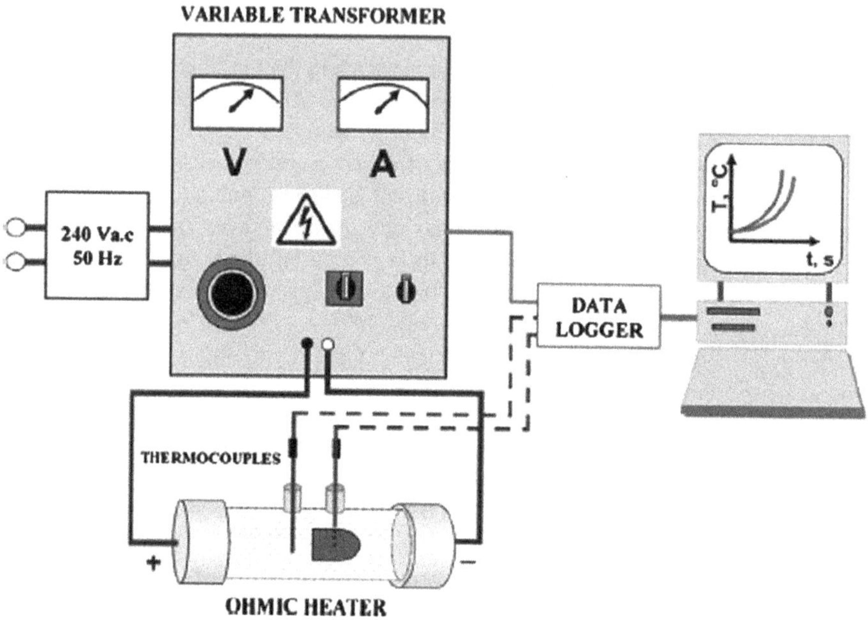

FIGURE 4.2 Principle of ohmic heating. *Source*: Lin (1998)

Ohmic blanching of milk and fruit juices was studied by Jakob et al. (2010) in order to compare inactivation of alkaline phosphatase, pectin methyl esterase and peroxidase with the results achieved by conventional indirect heating. The enhancement of enzyme inactivation rate in case of food materials due to the impact of ohmic heating was also observed. In addition, the peroxidase in vegetable juices was more prone to destabilization with ohmic heating.

4.4.4 Infrared Blanching

Infrared is electromagnetic radiation that falls between the regions of visible light waves (0.38–0.78 μm) and microwaves (1–1000 mm). Absorption of infrared by the rotation and vibration modes of water molecules in food generates the heat within the product (Rastogi, 2012). There are three types of infrared heater: near infrared (NIR) with wavelengths between 0.78 and 3 μm, mid infrared (MIR) with wavelengths between 3 and 50 μm, and far infrared (FIR) with wavelengths between 50 and 1000 μm. Foods having water, proteins, starches, and fats as their main constituents absorb far-infrared energy better than near infrared energy (Sakai and Hanzawa, 1994).

Infrared mainly heats the surface of the product rather than the immediate surroundings; the ambient temperature is therefore maintained at its normal level, making this technology energy efficient. It can be used in drying, baking, roasting, pasteurization, thawing, and blanching, and has the advantages of higher energy efficiency, shorter process time, and larger heat transfer coefficient. It also accommodates convective, conductive, and microwave heating. Table 4.4 shows applications of emerging and innovative blanching technologies.

TABLE 4.4
Applications of Emerging and Innovative Blanching Technologies

Blanching Technology	Product	Processing Conditions	Main Findings	References
Microwave blanching	Mushroom	Microwave heating at 85°C then water bath at 92°C for 20 s	PPO completely inactivated in 2 min, compared with conventional hot-water blanching requiring more than 6 min	Devece et al. (1999)
	Carrot slices	Microwave drying with 4 magnetrons (power of 1.25 kW for each magnetron) and a conveyer speed of 0.5 m/min	Nutritionally better than water and steam blanching.	Kidmose and Martens (1999)
	Mango, apple, orange, and banana	720 W for 1, 3 and 5 min	Preserves antioxidants and reduces browning by microwave blanching for 3 or 5 min	Romelle et al. (2020)
	Agaricus bisporus slices	2450 MHz microwave oven 800 W Galaxy Inc. Shunde, China 100°C/1 min	Slices with uniform microstructure and reduced time	Jiang et al. (2015)
	Star fruit	Microwave power levels 600, 480, 240 and 120 W for a slice thickness of 5, 10, and 15 mm and blanching time of 30–60, 40–70, 60–80, 80–150, and 300–720 s respectively.	Optimum power level 600 W with 30 s blanching time for 15 mm slice thickness	Behera et al. (2017)
	Red pepper	Power level 650, 750 and 900 W for 100 s	Residual activity values of PPO and POD are 9.80% and 16.43%, respectively; reduced drying time for 3.5 h, at power of 900 W for 100 s	Wang et al. (2017)

(Continued)

TABLE 4.4 (CONTINUED)
Applications of Emerging and Innovative Blanching Technologies

Blanching Technology	Product	Processing Conditions	Main Findings	References
Ohmic blanching	Strawberry	Heating at 60 Hz, 0–240 V, temperature at 30°C, 40°C and 50°C for 300 min, as the ratio of solution to fruit was 3:1 (w/w)	Greater mass transfer increases during osmotic dehydration (0.181–0.262) than unheated (0.149–0.223)	Moreno et al. (2013)
	Tomato juice	Electric field strength 24 V/cm, at 90°C for 15, 30, 45, and 60 s	PG (Polygalacturonase) and PME (Pectin methyl esterase) enzyme inactivation achieved by the OH (1 min.) was similar to that with CT (hot water bath of 90°C) of 5 min; the ascorbic acid of OH was 29–51% higher than CT; OH paste was more viscous and brighter red than CT treatment	Makroo et al. (2016)
	Blueberry pulp	Voltage (160, 172, 200, 228 and 240 V), at temperature of 90°C for 2 min	Lower degradation of anthocyanin (5.7–14.7%) than conventional heating	Sarkis et al. (2013)
	Apples	Temperature 30°C, 40°C and 50°C for 90 min, solution: fruit was 3:1 (w/w), with an alternating current at 60 Hz and 100 V, generating an electric field of 13V/cm	Combined with ohmic heating at 50°C was the best process for dehydrating apples	Moreno et al. (2013)
Infrared blanching	Carrot slices	Infrared radiation at 180–240°C for 8–15 min	Reduces moisture content by 13–23%; increases retention of vitamin C (62%), water (43%)	Vishwanathan et al. (2013)
	Potato chips	Effects of blanching and dehydrating using infrared (IR) radiation on the oil content of fried potato chips	PPO inactivates in 180 s and total oil content reduced by 13.79% compared to the conventional method.	Wu et al. (2017)

4.5 EFFECT ON FOOD QUALITY

Blanching reduces the carbohydrate, fat, and water-soluble content of some vegetables, but generally it helps to prevent fat-soluble nutrients from breaking down. It also causes some changes to the sensory and nutritional qualities of food.

Blanching facilitates the addition of sodium carbonate (0.125% w/w) or calcium oxide to water to protect chlorophyll and to retain the color of green vegetables, although the increase in pH may also increase losses of ascorbic acid. Apples and potatoes can be dipped in dilute (2% w/w) brine solution before blanching to prevent enzymatic browning.

Biscuits prepared from blanched pearl millet flour (steam blanching at 98°C for 30 s) contain a high level of calcium, iron, manganese, and phosphorus and their digestibility is also enhanced. In this case, blanching is important for leaching anti-nutrient compounds while maintaining nutritional and sensory qualities (Singh et al., 2006). It also causes physical and metabolic changes within food cells that result in cell death.

4.6 CONCLUSION

Blanching is an important unit operation widely used in many food processing industries, mainly to improve the quality of fruit and vegetables prior to canning, freezing, drying, and other processing techniques. It not only improves such product attributes as color, texture, and aroma, but also inactivates PPO and peroxidase (POD) enzymes, decreases the microbial load, reduces undesirable substances, and improves extractability of bioactive compounds, resulting in improved sensory and nutritional quality.

The water and steam blanching carried out in many food processes require high amounts of energy, cause nutritional loss from products, and generate waste water which contributes to water pollution. Emerging thermal blanching technologies, including microwave, ohmic, and infrared blanching therefore have many advantages both for processing and for society. These blanching processes reduce nutrient loss, waste water generation, and manufacturing costs, enabling products of high nutritional quality to be produced.

REFERENCES

Akanbi, C.T., A.O. Olumese, K.A. Taiwo, A. Ojo, and B.A. Akinwande. 2003. Effect of blanching medium on drying and storage characteristics of pepper. *Nig. Drying Symp. Series.* 1: 95–107.

Al-Khuseibi, M.K., S.S. Sablani, and C.O. Perera. 2005. Comparison of water blanching and high hydrostatic pressure effects on drying kinetics and quality of potato. *Dry. Technol.* *23*: 2449–2461.

Alvarez-Jubete, L., J. Valverde, A. Patras, A.M. Mullen, and B. Marcos. 2014. Assessing the impact of high-pressure processing on selected physical and biochemical attributes of white cabbage (Brassica oleracea L. var. capitata alba). *Food Biopro. Tech.* 7(3): 682–692.

Alzamora, S., L. Gerschenson, S. Vidales, and A. Nieto. 2000. Structural changes in the minimal processing of fruits: Some effects of blanching and sugar impregnation. In: P. Fito, E.O. Rodriguez and G. Barbosa-Canovas (eds.), *Food Engineering*. New York: Chapman and Hall, pp. 117–139.

Ando, Y., Y. Maeda, K. Mizutani, N. Wakatsuki, S. Hagiwara, and H. Nabetani. 2016. Impact of blanching and freeze-thaw pretreatment on drying rate of carrot roots in relation to changes in cell membrane function and cell wall structure. *J. LWT – Food Sci. Technol.* 71: 40–46.

Assiry, A., S.K. Sastry, and C. Samaranayake. 2003. Degradation kinetics of ascorbic acid during ohmic heating with stainless steel electrodes. *J. Appl. Electrochem.* 33: 187–196.

Bai, J.W., D.W. Sun, H.W. Xiao, A.S. Mujumdar and Z.J. Gao. 2013. Novel high humidity hot air impingement blanching (HHAIB) pretreatment enhances drying kinetics and color attributes of seedless grapes. *J. Innov. Food Sci. Emerg. Technol.* 20: 230–237.

Bajwa, U. and K.S. Sandhu. 2014. Effect of handling and processing on pesticide residues in food – A review. *J. Food Sci. Technol.* 51: 201–220.

Bastias, J.M., J. Moreno and C. Pia. 2015. Effect of ohmic heating on texture, microbial load, and cadmium and lead content of Chilean blue mussel (*Mytiluschilensis*). *Innov. Food Sci. Emerg. Technol.* 30: 98–102.

Behera, G., R. Kalpana, and K.N. Prakash. 2017. Effect of microwave blanching on slice thickness and quality analysis of star fruit. *J. Curr. Res. Nutr. Food Sci.* 5(3): 1–8.

Binsi, P.K., G. Ninan, A.A. Zynudheen, R. Neethu, V. Ronda, and C.N. Ravishankar. 2014. Compositional and chill storage characteristics of microwave blanched sutchi catfish *(Pangasianodonhypophthalmus)* fillets. *Int. J. Food Sci. Technol.* 49: 364–372.

Blasco, M., J.V. García-Pérez, J. Bon, J.E. Carreres, and A. Mulet. 2006. Effect of blanching and air flow rate on turmeric drying. *Food Sci. Technol. Int.* 12(4): 315–323.

Bonnechere, A., V. Hanot, R. Jolie, M. Hendrickx, C. Bragard, and T. Bedoret. 2012. Effect of household and industrial processing on levels of five pesticide residues and two degradation products in spinach. *J. Food Control.* 25: 397–406.

Boyes, S., P. Chevis, J. Holden and C. Perers. 1997. Microwave and water blanching of corn kernels: Control of uniformity of heating during microwave blanching. *J. Food Process. Preserv.* 21: 461–484.

Castro, S.M., A.S. Jorge, A.L.D. Jose, D. Ivonne, V.L. Ann, S. Chantal, and H. Marc. 2008. Effect of thermal blanching and of high pressure treatments on sweet green and red bell pepper fruits (*Capsicum annuum L.*). *Food Chem.* 107: 1436–1449.

Ceylan, E., W. McMahon and D.M. Garren. 2017. Thermal inactivation of listeria monocytogenes and salmonella during water and steam blanching of vegetables. *J. Food Prot.* 80(9): 1550–1556. doi: 10.4315/0362-028X.JFP-16-517.

Chafer, M., C. Gonzalez-Martınez, B. Fernandez, L. Perez, and A. Chiralt. 2003. Effect of blanching and vacuum pulse application on osmotic dehydration of pear. *Food Sci. Technol. Int.* 9: 321–328.

Dauthy, M.E. 1995. *Fruit and Vegetable Processing*. Rome: Food and Agriculture Organization of the United Nations.

Del Valle, J.M., V. Aranguiz, and H. Leon. 1998. Effects of blanching and calcium infiltration on PPO activity, texture, microstructure and kinetics of osmotic dehydration of apple tissue. *J. Food Res. Int.* 31(8): 557–569.

Devece, C., J.N. Rodriguez-Lopez, L.G. Fenoll, J. Tudela, J.M. Catala, and E. de los Reyes. 1999. Enzyme inactivation analysis for industrial blanching applications: Comparison of microwave, conventional, and combination heat treatments on mushroom poly phenoloxidase activity. *J. Agric. Food Chem.* 47: 4506–4511.

Dorantes-Alvarez, L., E. Jaramillo-Flores, K. Gonzalez, R. Martinez, and L. Parada. 2011. Blanching peppers using microwaves. *Procedia Food Sci.* 1: 178–183.

Farahnaky, A., R. Azizi, and M. Gavahian. 2012. Accelerated texture softening of some root vegetables by ohmic heating. *J. Food Eng.* 113: 275–280.

Fernández, C., A.M. Dolores, and W. Canet. 2006. The effect of low-temperature blanching on the quality of fresh and frozen/thawed mashed potatoes. *Int. J. Food Sci. Technol.* 41: 577–595. doi: 10.1111/j.1365-2621.2005.01119.x.

Halden, K., A. Aap De , and P.J. Fryer. 1990. Changes in the electrical conductivity of foods during ohmic heating. *Int. J. Food Sci. Technol.* 25: 9–25.

Hihat, S., H. Remini, and K. Madani. 2017. Effect of oven and microwave drying on phenolic compounds and antioxidant capacity of coriander leaves. *Int. Food Res. J.* 24(2): 503–509.

Hirota, T., K. Tahata, Y. Inoue, and K. Nakagawa. 2000. Conditions for making dry spinach by heated-air drying. *Bulletin Hyogo Prefect. Agric. Ins.* 48: 54–57.

Huan, Z., Z. Xu, and W. Jiang. 2015. Effect of Chinese traditional cooking on eight pesticides residue during cowpea processing. *J. Food Chem.* 170: 118–122.

Jakob, A., J. Bryjak, H. Wojtowicz, V. Ieova, J. Annus, and M. Polakovic. 2010. Inactivation kinetics of food enzymes during ohmic heating. *Food Chem.* 123: 369–376.

Jiang, N., C. Liu, D. Li, and Y. Zhou. 2015. Effect of blanching on the dielectric properties and microwave vacuum drying behavior of *Agaricus bisporus* slices. *J. Innov. Food Sci. Emerg. Technol.* 30: 89–97.

Kidmose, U. and H.J. Martens. 1999. Changes in texture, microstructure and nutritional quality of carrot slices during blanching and freezing. *J. Sci. Food Agric.* 79: 1747–1753.

Kingsly, A.R.P., V.M. Balsubramaniam, and N.K. Rastogi. 2012. Influence of high-pressure blanching on polyphenoloxidase activity of peach fruits and its drying behavior. *Int. J. Food Prop.* 12(3): 671–680.

Kingsly, R.P., R.K. Goyal, M.R. Manikantan, and S.M. Ilyas. 2007. Effects of pretreatments and drying air temperature on drying behaviour of peach slice. *Int. J. Food Sci. Tech.* 42: 65–69.

Knirsch, M.C., C. Alves dos Santos, O. de Martins, A.A. Soares Vicent, and T.C. Vessoni Penna. 2010. Ohmic heating – A review. *Trends Food Sci. Technol.* 21: 436–441. doi: 10.1016/jtifs.2010.06.003.

Lin, S. and M.S. Brewer. 2005. Effects of blanching method on the quality characteristics of frozen peas. *J. Food Quality* 28: 350–360.

Lin, T.M., Durance, T.D., and C.H. Scaman. 1998. Characterization of vacuum microwave, air and freeze dried carrot slices. *Food Res. Int.* 31: 111–117.

Llano, K.M., A.S. Haedo, L.N. Gerschenson, and A.M. Rojas. 2003. Mechanical and biochemical response of kiwifruit tissue to steam blanching. *Food Res. Int.* 6: 767–775.

Makroo, H.A., N.K. Rastogi, and B. Srivastava. 2016. Enzyme inactivation of tomato juice by ohmic heating and its effects on physicochemical characteristics of concentrated tomato paste. *J. Food Process. Eng.* doi: 10.1111/jfpe.12464.

Mazza, G. 1983. Drying of carrots: Effect of pre-drying treatments on moisture transport and product quality. *J. Food Eng.* 18: 113–123.

Moreno, J., R. Simpson, N. Pizarro, C. Pavez, F. Dorvil, and G. Petzold. 2013. Influence of ohmic heating/osmotic dehydration treatments on polyphenoloxidase inactivation, physical properties and microbial stability of apples (*cv. Granny Smith*). *Innov. Food Sci. Emerg. Technol.* 20: 198–207.

Nilnakara, S., N. Chiewchan, and S. Devahastin. 2009. Production of antioxidant dietary fibre powder from cabbage outer leaves. *J. Food Bioprod. Process.* 87: 301–307.

Orak, H.H. 2006. Effect of different blanching methods on peroxidase and lipoxygenase activity of broccoli and comparison of some nutritional value at enzymes inactivation levels. *Acta Alimentaria.* 35(3): 259–268.

Orikasa, T. and L. Wu, T. Shina, and A. Tagawa. 2008. Drying characteristics of kiwifruit during hot air drying. *J. Food Eng.* 85: 303–308.

Ramesh, M.N., W. Wolf, D. Tevini, and G. Jung. 1999. Studies on inert gas processing of vegetables. *J. Food Eng.* 40: 199–205.

Rastogi, N. 2012. Recent trends and developments in infrared heating in food processing. *Crit. Rev. Food Sci. Nutr.* 52: 737–760.

Reis, F.R., M.L. Masson, and Y. Waszczynskyj. 2008. Influence of a blanching pretreatment on color, oil uptake and water activity of potato sticks, and its optimization. *J. Food Process. Eng.* 31: 833–852. doi: 10.1111/j.1745-4530.2007.00193.x.

Rocha, T., A. Lebert, and C. Marty-Audouin. 1993. Effect of pretreatments and drying conditions on drying rate and colour retention of basil (*Ocimumbasilicum*). *J. LWT – Food Sci. Technol.* 26: 456–463.

Romelle, F.D., P.A. Emmanuel, R.P. Ashwini, M.T. Quentin, and M.M.F. Carl. 2020. Effect of microwave blanching on antioxidant activity, phenolic compounds and browning behavior of some fruit peelings. *J. Food Chem.* 125: 302–308. doi: 10.1016/j.foodchem.2019.125308.

Sablani, S.S., P.K. Andrews, N.M. Davies, T. Walters, H. Saez, and L. Bastarrachea. 2011. Effects of air and freeze drying on phytochemical content of conventional and organic berries. *Drying Technol.* 29: 205–216.

Sakai, N. and T. Hanzawa. 1994. Applications and advances in far infrared heating in Japan. *Trends Food Sci. Technol.* 5: 357–362.

Sarkis, J.R., D.P. Jaeschke, I.C. Tessaro, and L.D.F. Marczak. 2013. Effects of ohmic and conventional heating on anthocyanin degradation during the processing of blueberry pulp. *LWT – Food Sci. Technol.* 51: 79.

Sastry, S.K. and J.T. Barach. 2000. Ohmic and inductive heating. *J Food Sci.* 65: 42–46.

Shivhare, U.S., M. Gupta, S. Basu, and G.S.V. Raghavan. 2009. Optimization of blanching process for carrots. *J. Food Process. Eng.* 32: 587–605.

Singh, G., S. Sehgal, and A. Kawatra. 2006. Mineral profile, anti-nutrients and in vitro digestibility of biscuit prepared from blanched and malted pearl millet flour. *Nutr Food Sci.* 36: 231–239. doi: 10.1108/00346650610676802.

Smith, P.G. 2011. *Introduction to Food Process Engineering*, 2nd edn. Heidelberg: Springer.

Teng, S.S. and B.H. Chen. 1999. Formation of pyrochlorophylls and their derivatives in spinach leaves during heating. *Food Chem.* 65: 367–373.

Thostenson, E.T. and T.W. Chou. 1999. Microwave processing: Fundamentals and applications. *Compos.* 30: 1055–1071.

Turhan, M.K.N. and F. Sahbaz. 1997. Drying kinetics of red pepper. *J. Food Proc. Pres.* 21: 209–223.

United States Environmental Protection Agency. 1974. *Continuous In-Plant Hot-Gas Blanching of Vegetables*. Washington, DC: USEPA.

Vighi, M. and A. Di Guardo. 1995. *Book: Pesticide Risk in Groundwater*. London: CRC Press, 288 pp. ISBN: 9780873714396.

Villalobos, M.C., M.J. Serradilla, and A. Martin. 2016. Evaluation of different drying systems as an alternative to sun drying for figs *(Ficuscarica L). Innov. Food Sci. Emerg. Technol.* 36: 156–165. doi: 10.1016/j.ifset.2016.06.006.

Vishwanathan, K.H., G.K. Giwari, and H.U. Hebbar. 2013. Infrared assisted dry-blanching and hybrid drying of carrot. *Food Bioprod. Process.* 91: 89–94.

Wang, J., X.H. Yang, A.S. Mujumdar, D. Wang, J.H. Zhao, and X.M. Fang. 2017. Effects of various blanching methods on weight loss, enzymes inactivation, phytochemical contents, antioxidant capacity, ultra structure and drying kinetics of red bell pepper (*Capsicum annuum L.*). *LWT – Food Sci. Technol.* 77: 337–347.

Wang, N., M.J. Lewis, J.G. Brennan, and A. Westby. 1997. Effect of processing methods on nutrients and anti-nutritional factors in cowpea. *Food Chem.* 58: 59–68.

Watanabe, T., T. Orikasa, K. Sasaki, S. Koide, T. Shiina, and A. Tagawa. 2014. Influence of blanching on water transpiration rate and quality changes during far-infrared drying of cut cabbage. *J. Japanese Soc. Agric. Mach. Food Eng.* 76: 387–394.

Wu, B., J. Wang, Y. Guo, Z. Pan, and H. Ma. 2017. Effects of infrared blanching and dehydrating pretreatment on oil content of fried potato chips. *J. Food Process. Preserv.*, pp. 1–7. doi: 10.1111/jfpp.13531.

Yoshida, H., T. Orikasa, S. Koide, Y. Muramatsu, and A. Tagawa. 2014. Application of brief hot water soaking as a pre-treatment in kiwifruit drying. *J. Japanese Soc. Food Sci. Technol.* 61: 151–159.

Zhu, Y. and Z. Pan. 2009. Processing and quality characteristics of apple slices under simultaneous infrared dry-blanching and dehydration with continuous heating. *J. Food Eng.* 90: 441–452.

5 Thermal Processing of Food

K.K. Dash

Department of Food Processing Technology, Ghani Khan Choudhury Institute of Engineering and Technology, Malda, West Bengal, India

Pinki Deka

Department of Food Engineering and Technology, Tezpur University, Tezpur, Assam, India

CONTENTS

5.1 INTRODUCTION

Biological materials, particularly foods, are perishable. It is therefore essential to apply certain preservation techniques in order to increase their shelf life. Physical and chemical preservation methods like drying, smoking, salting, chilling, freezing, and heating are the most commonly used in food industries. Thermal processing causes surface destruction of pathogenic and/or spoilage microorganisms, leading to increased shelf life for products. Sterilization refers to any process that eliminates or kills all forms of life, including transmissible agents (such as fungi, bacteria, viruses, spore forms, etc.) present on a surface, contained in a fluid, in medication, or in a compound such as biological culture media. Sterilization can be achieved by applying heat, chemicals, irradiation, high pressure, and filtration or combinations thereof. Generally, food industries use *Clostridium botulinum* as the index microorganism; in other words, complete sterility with respect to this spore is essential. The sterilization process can be divided into two main categories: in-container, which is used for foods placed in containers such as cans, bottles, and plastic pouches; and the continuous flow system for ultra-heat treatment (UHT) processes, which generally involves heating at 140–150°C for 1–3 s. Sterilization by definition terminates all life, whereas sanitization and disinfection terminate selectively and partially. Both sanitization and disinfection reduce the number of targeted pathogenic organisms to what are considered 'acceptable' levels – levels that a reasonably healthy, intact body can deal with.

It is widely known that temperature influences the growth rate of microorganisms, and each organism has a certain optimal temperature range at which it grows best. Exposure beyond this temperature for relatively long duration causes microbial death.

5.2 THERMAL DEATH RATE KINETICS OF MICROORGANISMS

When microorganisms are subjected to temperatures beyond their optimum, they lose their viability. Assuming that inactivation of a single enzyme will inactivate the cell, and then in a suspension of microorganism of a single species at a constant temperature, the death rate, i.e., the rate of destruction, is proportional to the number of organisms:

$$\frac{dN}{dt} = -kN \tag{5.1}$$

The above is a first-order differential equation where

N = the number of viable organisms at a given time,

t = time (min), and

k = reaction velocity constant (min^{-1}) which is a constant for a particular temperature and the type of organism.

Rearranging and integrating Equation (5.1), Equations (5.2)–(5.5) were obtained:

$$\int_{N_0}^{N} \frac{dN}{N} = -\int_{t=0}^{t} k\,dt \tag{5.2}$$

$$\ln \frac{N_0}{N} = kt \tag{5.3}$$

$$2.303 \log \frac{N_0}{N} = kt \tag{5.4}$$

$$\log \frac{N_0}{N} = \frac{kt}{2.303} \tag{5.5}$$

where N_0 = the original number of microorganisms at t = 0, also called the concentration level (initial concentration of microbes) and N = the number of microorganisms at time t, also called the sterility level.

5.3 DECIMAL REDUCTION TIME (D-VALUE)

D-value (D-Value = Time) refers to decimal reduction time and is the time required at a certain temperature to kill 90% of the organisms being studied. Hence decimal reduction time is the time (in min) required to achieve a 10-fold reduction in viability of a microbial population at a given temperature under standard conditions.

Substituting t by D into Equation (5.3), Equation (5.6) was obtained:

$$\frac{N}{N_0} = \frac{1}{10} = e^{-kD} \tag{5.6}$$

Or, taking $\log_{10}$ on both sides and solving for D, Equation (5.7) was obtained:

$$D = \frac{2.303}{k} \tag{5.7}$$

Combining Equations (5.5) and (5.7) the following was obtained:

$$\log_{10} \frac{N_0}{N} = \frac{t}{D} \tag{5.8}$$

A plot of $\log_{10}$ (N/N_0) vs. t results in a straight line (Equation 5.5). It was observed that sometimes a bacterial spore plot deviates somewhat from the logarithmic rate of death, particularly during the short period immediately following thermal processing. However, a logarithmic-type curve was generally used for microbial spores. To determine the microbial death rate, the cell suspension was sealed in a test tube which was then subjected to a high temperature in a hot-water bath. After a specific duration of time, it was removed and immediately chilled. The number of viable organisms was then determined by the standard plate count method. Figure 5.1 shows the effect on an organism held at specified temperature for a period of time.

At time t = 20 min, the number of microorganisms was 10^7 and after 10 min, 90% of them had been destroyed, leaving 10^6 microorganisms. This is referred to as a

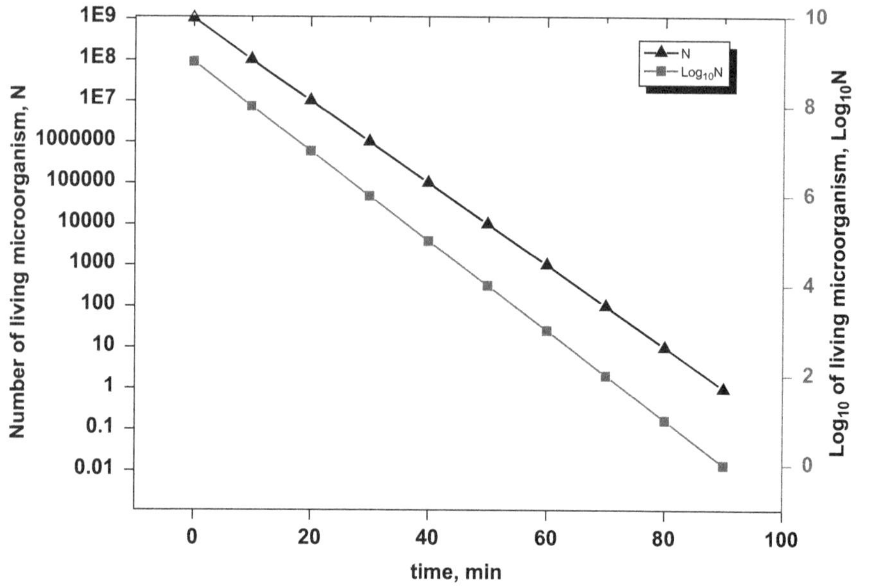

FIGURE 5.1 Decimal reduction time

one-decimal log reduction. In this instance D-value = 10 min. For every additional 10 min at the same temperature, this process kills an additional 90% of the organisms present. The D-value for an organism is significantly different at different temperatures. D-value changes dramatically when the temperature is varied. It is easy to understand that, as the temperature increases, it takes less time to destroy a population of organisms. Therefore, the D-value (the time needed to kill 90% of the population) becomes smaller as the temperature rises. The decimal reduction time is dependent on the temperature, the type of microorganism, and the composition of the medium containing the microorganism. It is worth noting that D-value remains the same for each log cycle and does not depend on the initial concentration of microorganisms (Equation 5.7).

Example 1:

A thermal resistance experiment resulted in a D-value of 7.5 min at 110°C. If there were 4.9×10^4 survivors at 10 min, determine the ratio (N/N_0) for 5, 10, 15, and 20 min.

Solution:

D = 7.5 min
Therefore, k = 2.303/7.5 = 0.307

$$N / N_0 = e^{-kt}$$

TABLE 5.1
Ratio (N/N_0) at Different Time Intervals

Time, min	N/N_0
5	= exp. (−0.307×5) = 0.2154
10	= exp. (−0.307×*10) = 0.0464
15	= exp. (−0.307×*15) = 0.01
20	= exp. (−0.307×*15) = 0.0021

The values of the ratio N/N_0 obtained at different time interval is presented in Table 5.1.

5.4 EVALUATION OF THE EFFECT OF TEMPERATURE ON THERMAL INACTIVATION

The D- and k-values are temperature dependent, and the Arrhenius equation describes the temperature dependence of the parameters. The effect of temperature on the reaction rate k may be expressed by an Arrhenius-type equation as shown in Equations (5.9)–(5.11).

$$k = A e^{-E_a/RT} \tag{5.9}$$

$$\ln k = \ln A - \frac{E_a}{RT} \tag{5.10}$$

$$\ln \frac{k}{A} = -\frac{E_a}{RT} \tag{5.11}$$

where A = an empirical constant known as Arrhenius Constant, R is the universal gas constant, i.e., 8.314 J/(mol·K), T is absolute temperature in K, and E_a is the activation energy in kJ/gmol. The activation energy value ranged from 210 to 418 kJ/gmol for vegetative cells and spores and much less for enzymes and vitamins. It should be evident that a plot of ln (k) versus 1/T will be a straight line with a negative slope ($-E_a/R$), i.e., ln k decreases as temperature decreases. Based on the Arrhenius model (Equation 5.9), the slope of In(k) versus 1/T gives the activation energy. The activation energy constant describes the influence of temperature on the magnitude of the first-order reaction rate constant (k).

The logarithmic relationship between the rate of thermal inactivation and the temperature is expressed in Equation (5.12).

$$\ln \frac{k_1}{k_2} = -\frac{E_a}{R}\left[\frac{1}{T_1} - \frac{1}{T_2}\right] \tag{5.12}$$

$$\ln \frac{k_1}{k_2} = -\frac{E_a}{R}\left[\frac{T_2 - T_1}{T_1 T_2}\right] \tag{5.13}$$

Similarly, the temperature dependence of the D-value in terms of the activation energy is presented in Equation (5.14).

$$\ln\frac{D_1}{D_2} = -\frac{E_a}{R}\left[\frac{1}{T_1} - \frac{1}{T_2}\right] \tag{5.14}$$

5.5 Z-VALUE OR TEMPERATURE COEFFICIENT OF MICROBIAL DESTRUCTION

The z-value is defined as the temperature coefficient of microbial destruction, i.e., z-value is the number of degrees of temperature increase necessary to reduce the D-value one log cycle, that is, to kill 90% of the population of a microorganism (Figure 5.2). For z-values, temperature is the variable, with time being constant. In Figure 5.3 the D-value varied by 10 times for a 10°C variation in temperature, when z = 10.

Let us assume that the D-value of a specific microorganism at 131°C is 50 s and the z-value is 10°. If the heating temperature was increased by 10°C (141°C), the D-value would decrease 10-fold to 5 s. Similarly, if heating temperature was decreased by 10°C (121°C), the D-value would increase 10-fold to 500 s.

The z-value is the reciprocal of the slope resulting from the plot of the logarithm of the D-value versus the temperature at which the D-value was obtained. While the D-value gives the time needed at a certain temperature to kill an organism, the z-value expresses the resistance of an organism to differing temperatures. For sterilization processes using steam, the z-value varies between 6 and 13 within the temperature range of 100–130°C. In the traditional approach to thermal processing, the term thermal resistance constant was used for the z-value to describe the influence of

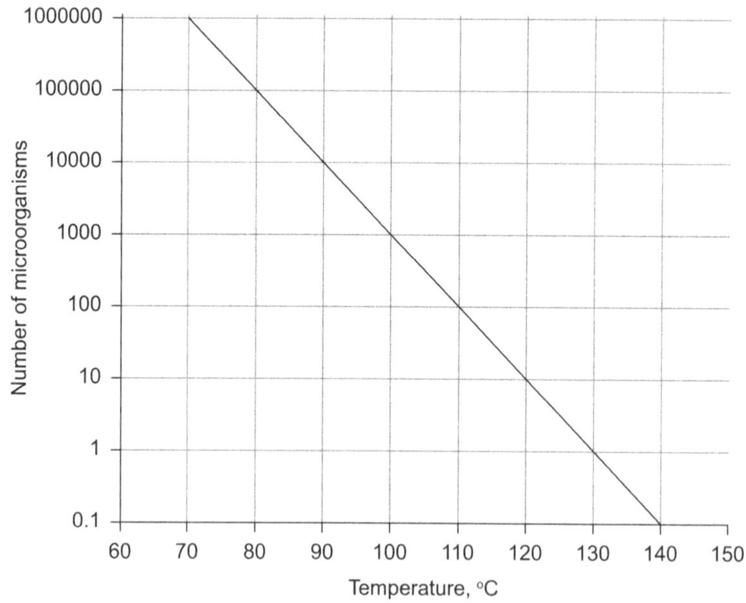

FIGURE 5.2 Plot of number of microorganisms with respect to temperature

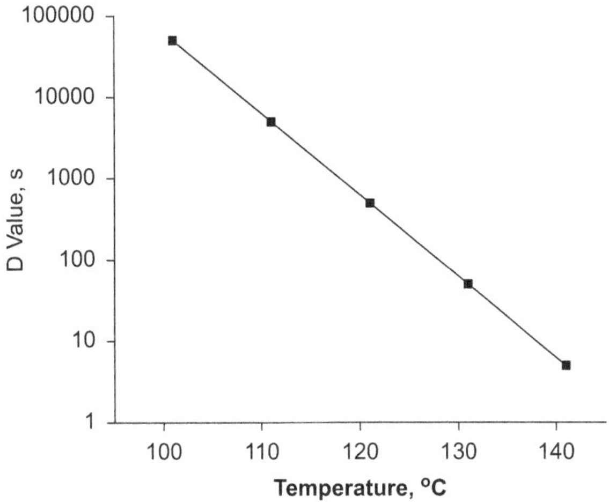

FIGURE 5.3 Plot showing z-value and D-value

temperature on decimal reduction time for microbial population. The z-value can be expressed as presented in Equations (5.15)–(5.18).

$$\log \frac{D_{Tref}}{D_{T2}} = \frac{T_2 - T_{ref}}{z} \tag{5.15}$$

$$z = \frac{T_2 - T_{ref}}{\log D_{Tref} - \log D_{T2}} \tag{5.16}$$

$$\frac{D_{T2}}{D_{Tref}} = 10^{\frac{T_{ref} - T_2}{z}} \tag{5.17}$$

$$D_{T2} = D_{Tref}\, 10^{\frac{T_{ref} - T_2}{z}} \tag{5.18}$$

Example 2:

In an experiment, the z-value for *Campylobacter jejuni* spores in whole meat at 121°C was found to be 12°C and the decimal reduction time at corresponding temperature was 45 s. Find the D-value for the same at 150°C.

Solution:

Data given:

$$D_{Tref} = 45\,s, T_{ref} = 121°C, T = 150°C, z = 12°C$$

$$D_{T2} = D_{Tref}\, 10^{\frac{T_{ref} - T_2}{z}}$$

$$D_{150} = 45 \times 10^{\frac{121-150}{12}} = 0.175\,s$$

5.6 THERMAL DEATH TIME (F-VALUE)

Thermal death time is the amount of time required to kill a specific number of microbes at a specific temperature. This value is obtained by keeping the temperature constant and measuring the time necessary to kill the amount of cells specified, which is expressed as a multiple of D-values as long as the survivor curve follows a first-order model. Effective sterilization kills or destroys the target microorganisms, infectious proteins, viruses, etc. Increasing the time and/or temperature of sterilization can result in the complete elimination of microbes. The F-value of a process is the time needed to destroy a known number of microorganisms at any reference temperature and z-value. This means that, within a certain sterilization phase, the F-value can be expressed in minutes at a certain temperature T_2 as presented in Equation (5.19).

$$F_{T2} = F_{Tref}\, 10^{\frac{T_{ref} - T_2}{z}} \tag{5.19}$$

The lethal rate was calculated using the formula in Equation (5.20).

$$\text{Lethal rate} = 10^{\frac{T_{ref} - T_2}{z}} \tag{5.20}$$

where T_2 is the temperature, in Celsius, at which the lethal rate is required and T_{ref} is the reference temperature

The total F-value (lethal value) of a process takes account of the heating-up and cooling-down phases of the cycle and can be calculated by integration of lethal rates with respect to time at discrete temperature intervals. For several different temperature stages, each having different times t_1, t_2, …., the F values for each stage are added to give the total F_o.

$$\mathrm{F} = t_1.10^{(\mathrm{T1}-250)/z} + t_2.10^{(\mathrm{T2}-250)/z} + \ldots \tag{5.21}$$

The F_0 value of a process is the lethality expressed in terms of the equivalent time in minutes at a temperature of 121°C delivered by the process to the product in its final container with reference to microorganisms possessing a z-value of 10°C. Under actual conditions of use the autoclave heats the substance for a certain period of time before the sterilization temperature, i.e., 121°C, is reached. Microorganisms are killed during the heating time (up to 121°C) and during cooling time. The residence time can be correspondingly shortened considering the heating and cooling cycle during the process. This allows the F-value to be achieved in a shorter overall process time. Substituting T_1 = 121°C (250°F), which is the standard temperature against which thermal processes are compared, and replacing $F_{T2} = F_T$ and T_2 = T, Equation (5.18) can be presented as shown in Equation (5.22).

$$F_T = F_{121} 10^{(121-\mathrm{T})/z} \tag{5.22}$$

Example 3:

When same lethality as required for F_{121} of 6.75 min is delivered, determine the two equivalent processes at 105°C and 150°C where the z-value is 12°C.

Solution:

$$F_T = F_{Tref} \times 10^{(Tref-T)/z}$$

$$F_{105} = 6.75 \times 10^{(121-105)/12} = 144.3 \text{ min}$$

$$F_{150} = 6.75 \times 10^{(121-150)/12} = 0.026 \text{ min}$$

Example 4:

A thermal process is accomplished by instantaneous heating to 130°C followed by a 6-sec hold and immediate cooling. Estimate the lethality at 121°C when the thermal resistance for the microorganism is 7.5°C.

Solution:

$$\frac{F_{130}}{F_{121}} = 10^{(130-121)/7.5} = 15.848$$

$$F_{130} = F_{121} \times 15.848 = 95.1 \text{ s.}$$

Example 5:

Let us assume the decimal reduction time of *L. monocytogenes* spore for pasteurization of apple juice at temperature 90°C is 120 s. The z-value of the spore is 11 s. (a) Evaluate the decimal reduction time of the spore at temperature 68°C and 112°C. (b) Evaluate the time required for 12 log cycles reduction at 68°C and 112°C.

Solution:

$$\frac{D_{68}}{D_{90}} = 10^{(90-68)/11} = 100$$

$$D_{68} = 100 \times 120 = 12000 \text{ s}$$

$$\frac{D_{112}}{D_{90}} = 10^{(90-112)/11} = 0.01$$

$$D_{112} = 0.01 \times 120 = 1.2 \text{ s}$$

Time required for 12 log cycles reduction at 68°C is

$$F_{68} = 12D_{68} = 12 \times 12000 = 144000 \text{ s}$$

Time required for 12 log cycles reduction at 112°C is

$$F_{112} = 12D_{112} = 12 \times 1.2 = 14.4 \text{ s}$$

Example 6:

Mango juice was subjected to a sterilization process for *Alicyclobacillus acidoterrestris* microorganism. The initial count of the microorganism in the food was 2×10^4/g and the container has a capacity of 500 g. Given that D_{110} = 3.55 min and z-value was 9.5°C, a process that will ensure spoilage occurs in less than one container per 100, 000 is desired. Calculate the minimum holding time required to get the desired outcome. The temperatures at selected points are as follows:

Time (s)	0.5	1	1.5	2	3	4	5	6
Temperature (°C)	82.16	88.54	94.91	101.29	107.67	114.05	120.43	126.80

Time (s)	7	10	11	12	13	14	15	16
Temperature (°C)	130.00	130.00	122.21	114.08	105.96	97.83	89.70	85.89

Solution:

Step 1:

Initial count of the microorganisms in the food = $2 \times 10^4\ g^{-1}$

Final count of the microorganisms in the food = $\dfrac{1}{500 \times 10^5}\ g^{-1}$

Number of reduction cycles = $\dfrac{1}{500 \times 10^5 \times 2 \times 10^4}$ = 12 log cycles

Step 2:

Calculation of death time (F) at 110°C and 130°C

$F_{110°C} = 12D_{110°C} = 12*3.55 = 42.60$ min

Calculate the equivalent thermal death time (F) at 110°C

$F_{130°C} = F_{110°C}*10^{(110-130)/9.5}$

$F_{130°C} = 0.00784*42.60 = 0.3343$ s = 20.05 s

Lethal rate = $10^{(T1 - 130)/z}$

Lethality = $\Delta t.\ 10^{(T1 - 130)/z}$

	Time (s)	Temperature (°C)	Δ t(s)	Average tempera-ture (°C)	Lethal rate	Lethality (s)	Accumulated lethality (s)
Heating	0.5	82.16					
Heating	1	88.54	0.5	85.34715	1.99388×10^{-5}	9.96942×10^{-6}	9.96942×10^{-6}
Heating	1.5	94.91	0.5	91.72525	9.35576×10^{-5}	4.67788×10^{-5}	5.67482×10^{-5}

	Time (s)	Temperature (°C)	Δ t(s)	Average temperature (°C)	Lethal rate	Lethality (s)	Accumulated lethality (s)
Heating	2	101.29	0.5	98.10335	0.000438993	0.000219497	0.000276245
Heating	3	107.67	1	104.4815	0.002059856	0.002059856	0.002336101
Heating	4	114.05	1	110.8596	0.00966531	0.00966531	0.012001411
Heating	5	120.43	1	117.2377	0.045351822	0.045351822	0.057353233
Heating	6	126.80	1	123.6158	0.21280101	0.21280101	0.270154242
Heating	7	130.00	1	128.4024	0.678940286	0.678940286	0.949094528
Holding	10	130.00	3	130	1	3	3.949094528
Cooling	11	122.21	1	126.1055	0.389092296	0.389092296	4.338186824
Cooling	12	114.08	1	118.1477	0.056544384	0.056544384	4.394731207
Cooling	13	105.96	1	110.0211	0.007887836	0.007887836	4.402619044
Cooling	14	97.83	1	101.8945	0.001100339	0.001100339	4.403719382
Cooling	15	89.70	1	93.7679	0.000153495	0.000153495	4.403872878
Cooling	16	85.89	1	87.7973	3.61086×10^{-5}	3.61086×10^{-5}	4.403908986

Step 3: Additional holding time and total holding time requirement

Additional holding time required at 130°C = 20.05–4.40 = 15.65

Holding time maintained in the above process = 3 s

Total minimum holding time required to get the desired outcome = 15.65 + 3 = 18.65 s

The temperature time plot, lethal rate and time plot, and accumulated lethality and time plot for sterilization of mango juice are presented in Figures 5.4, 5.5, and 5.6.

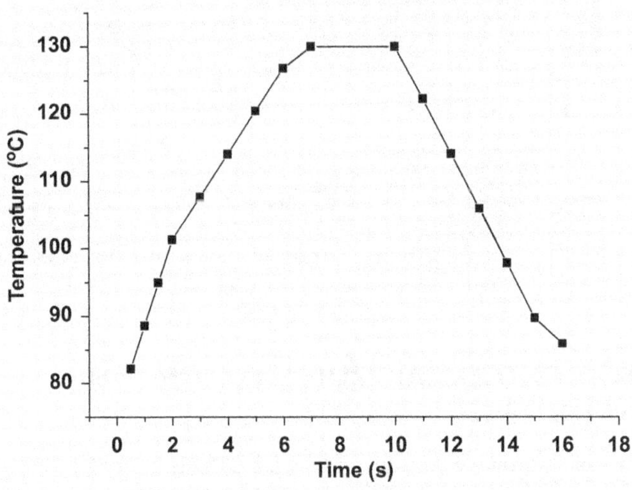

FIGURE 5.4 Temperature and time plot for sterilization of mango juice

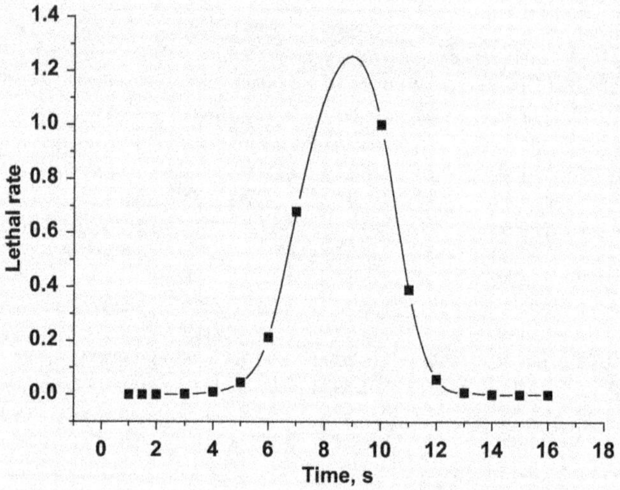

FIGURE 5.5 Lethal rate and time plot for sterilization of mango juice

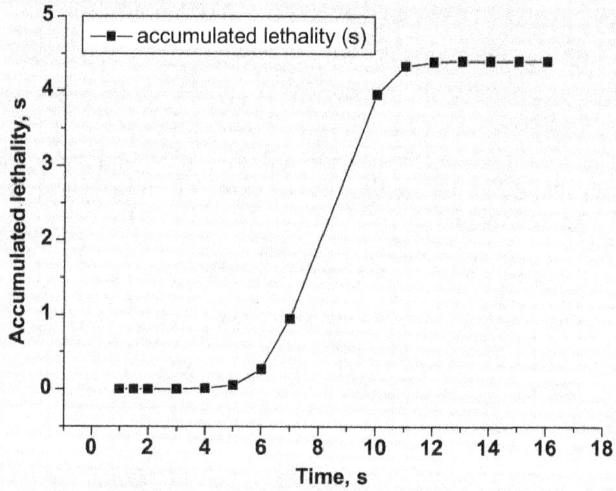

FIGURE 5.6 Accumulated lethality and time plot for sterilization of mango juice

5.7 EFFECTS OF THERMAL PROCESSING ON FOOD CONSTITUTENTS

Thermal processing produces certain undesirable effects in food, primarily the reduction of certain nutritional values. For example, certain heat-labile constituents like ascorbic acids, thiamin and riboflavin (Vitamin B_1 and B_2) are partially destroyed by thermal processing. Thermal processing also contributes to flavor change in a food product and this same kinetics can be applied to predict the time for detecting a flavor

TABLE 5.2
D-value, z-value, and Activation Energy of Different Thermal Processing Operations

Nutrient in Product	Ea(kJ/mol)	D (min)	Z (°C)	References
Vitamin C in orange juice	65	$46_{T = 100\,^\circ C}$	19.2	Vikram et al. (2005)
Anthocyanin in raspberries	75	$112_{T = 110\,^\circ C}$	37.4	Verbeyst et al. (2011)
Anthocyanin in purple potato	73	$32_{T = 120\,^\circ C}$	47.9	Nayak et al. (2011)
Ascorbic acid in pink guava pulp	95.82	$39.7_{T = 85\,^\circ C}$	23.5	Vishwasrao and Ananthanarayan (2018)
Lycopene pink guava pulp	54.92	$78.3_{T = 85\,^\circ C}$	41.7	Vishwasrao and Ananthanarayan (2018)
Pomelo fruits (OH (60 Hz))	1535	$2.16_{T = 52\,^\circ C}$	4.6	Khue et al. (2020)
Pomelo fruits (OH (500 Hz))	1411.9	$1.83_{T = 52\,^\circ C}$	4.9	
Pomelo fruits (Conven.)	385	$2.10_{T = 52\,^\circ C}$	5.3	
Polyphenol oxidase in Gold pear	375	$3.3_{T = 85\,^\circ C}$	6	Sulaiman et al. (2015)
Royal Gala apple	134	$2_{T = 85\,^\circ C}$	17	
Camarosa strawberry	147	$16_{T = 85\,^\circ C}$	14	

change in a product. Table 5.2 shows the D-value, z-value, and activation energy of various phytochemical compounds, such as vitamin C, anthocyanin, lycopene, and enzymes (polyphenol oxidase) in thermal processing of different food products. The D-values and z-values of various microorganisms for thermal processing of food products are presented in Table 5.3.

5.8 THERMAL STERILIZATION

Sterilization is one of the most widely used techniques in all the food processing industries. Sterilization is defined as statistically complete destruction of all microorganisms, including the bacteria and spores that are most resistant. The different physical methods of microbial control for food preservation include the application of high temperatures, radiation, filtration, and desiccation (drying). Many of these approaches destroy cells by denaturation, decay, or chemical alteration through destroying membranes, increasing membrane permeability, or damaging proteins and nucleic acids. Sterilization in food is done at a temperature between 121 and 140°C (Lin, 1976; Deak, 2014). Heating is one of the most well-known and well-established types of microbial control, used in simple techniques like cooking and canning. Heat can kill microbes by altering their membranes and denaturing proteins.

Escherichia coli O157:H7 is an anaerobic microorganism which is facultative in nature. It has been seen that *E. coli* O157:H7 is responsible for hemorrhagic colitis and also shows hemolytic uremic syndrome (Tarr, 1994). Experiments were conducted to find D-values in lean beef and chicken using linear regression and by a survival equation. The survival equation was a model that was fitted to the non-linear survival curve to obtain two D-values. The changes in *E. coli* O157:H7 heat resistance were assessed in beef during refrigeration as well as in frozen storage (Juneja

TABLE 5.3
D-values and z-values of Different Microorganisms for Thermal Processing of Food Products

Microorganism	Product	D (min)	Z (°C)	References
Alicyclobacillus acidoterrestris	Apple juice	$11.10_{T=90°C}$	8.50	Bahceci and Acar (2007)
	Apple juice	$41_{T=80°C}$	12.2	Komitopoulou et al. (1999)
	Apple juice	$56_{T=85°C}$	7.7	Splittstoesser et al. (1994)
	Berry juice	$11_{T=88°C}$	7.2	Walls (1997)
	Cupuacu extract juice	$17.5_{T=85°C}$	9.0	Silva et al. (1999)
	Mango pulp juice	$40_{T=80°C}$	21.27	De Carvalho et al. (2008)
	Tangerine juice	$15_{T=90°C}$	10.8	Lopez et al. (2011)
E. coli O157:H7	Apple juice	$4.1_{T=56°C}$	5.6	Mazzotta (2001)
	Orange juice	$7.5_{T=56°C}$	4.8	
	White grape juice	$4_{T=56°C}$	5.7	
L. monocytogenes	Apple juice	$1.59_{T=56°C}$	6.3	Mazzotta (2001)
	Orange juice	$2.05_{T=56°C}$	6.0	
	White grape juice	$2.29_{T=56°C}$	6.6	
Salmonella	Apple juice	$0.96_{T=56°C}$	6.0	Mazzotta (2001)
	Orange juice	$1.61_{T=56°C}$	4.6	
	White grape juice	$2.43_{T=56°C}$	6.2	
Alicyclobacillus acidoterrestris	Orange juice (200 MPa)	$5.0_{T=65°C}$	21.3	Silva et al. (2012)
	Orange juice (600 MPa)	$3.4_{T=65°C}$	34.4	
Alicyclobacillus acidotexrrestris (E14)	Passion fruit juice	$1.4_{T=95°C}$	6.9	McKnight et al. (2010)
Alicyclobacillus acidotexrrestris (E25)	Passion fruit juice	$1.8_{T=95°C}$	7.9	McKnight et al. (2010)

et al., 1997). The study was conducted with four strains of *E. coli* O157:H7 – EDL-931, 45753–35, Cl-9218, and 933. Approximately 0.1 mL of dilution of *E. coli* O157:H7 were inoculated into 95% of lean ground beef and lean ground chicken samples, to give a final cell concentration of 7 $\log_{10}$ CFU/g. To study the heat resistance nature of *E. coli* O157:H7 during refrigerated and frozen storage, bags with meat samples were frozen at −18°C and refrigerated at 4°C for 48 h. They were submerged in a water bath, the temperature of which was controlled to a range of 55–65°C. Readings from a thermocouple inserted into the meat pouches were taken every second including come-up time. The 90% lean ground beef and ground chicken had a pH of 6. It was found that fresh chicken when subjected to heat at 60°C showed a linear reduction of population but beef inoculated with *E. coli* O157:H7 showed a deviation from log-linear decline in the population with time. The D-values of *E. coli* O157:H7 in ground beef and chicken at temperatures of 55, 57.5, 60, 62.5, and 65°C were calculated.

The D-values of *E. coli* O157:H7 calculated using linear regression in beef ranged between 21.13 min (55°C) and 0.39 min (65°C). Using a survival model for non-linear survival curves, D-values in beef ranged from 20.45 min (55°C) to 0.16 min (65°C). The z-values of beef and chicken were determined and found to range between 4.94 and 6.79°C. In the study of the effect of storage and refrigeration on the heat resistance of *E. coli* O157:H7, it was found that in fresh ground beef when subjected to heating at 60°C, there was a reduction in the cells of *E. coli* O157:H7 by 2.48 logs (from 7.69 $\log_{10}$ CFU/g to 5.21 $\log_{10}$ CFU/g) within 2 min and 6.7 logs at 16 min. When applied to refrigerated and frozen meat samples, i.e., beef at 60°C for 2 min, there was a reduction by 0.57 and 0.21 log in *E. coli* O157:H7 colony counts/g from its initial concentration of 7.85 and 7.56 $\log_{10}$ CFU/g respectively. When applied for 16 min, the log reduction was found to be 5.65 and 5.05 $\log_{10}$ CFU/g respectively (Juneja et al., 1997). The thermal death time values are useful in the design of food-processing equipment with acceptable critical control points to provide complete protection against *E. coli* O157:H7 in cooked beef and chicken.

5.9 THERMAL PASTEURIZATION

Pasteurization is the process in which foods are subjected to heat treatment below 100°C in order to destroy microorganisms that are of public health concern. Pasteurization does not cause complete killing or removal of microorganisms in foods but aims at killing pathogens and lowering spoilage microorganisms that start to grow during storage and distribution (Silva and Gibbs, 2010). The use of pasteurization to kill pathogens in food in order to increase shelf life has become widespread.

Pectin esterase (PE) is responsible for cloudiness and spoilage of citrus juices during storage. Orange juice was subjected to pasteurization treatment to enhance its shelf life by preventing cloudiness and subsequent spoilage (Chen et al., 1993). Thermal pasteurization of orange juice based on the thermal destruction characteristics of both forms of PE, i.e., thermostable and thermolabile forms, was analyzed (Atkins and Rouse, 1953; Eagerman and Rouse, 1976). The log linear kinetic model conventionally used was not a good fit for the pasteurization of citrus juice. Lee et al. (2003) therefore conducted a study to design an improved non-linear model to represent the pasteurization process.

The residual activity of PE in Valencia orange juice was evaluated when juice was subjected to heat within a temperature range of 75–90°C to find the reliability of the kinetic model of the two-component system. In addition, the D- and z-values of PE in the orange juice sample were determined. Orange juice at a concentration of 12.3^0 B and pH of 3.6 was subjected to rapid heating and cooling with four different temperatures – 75, 80, 85, and 90°C – with varying holding times (30 s to 30 min). The PE activity was determined using the titrimetric method (Barrett and Gonzalez, 1994). Five replicates of samples were averaged to represent each time–temperature combination for the treatment.

The study showed that with increase in temperature, the residual PE activity reduced. Residual PE activity was 0.1% when the juice was subjected to 90°C for 60 s. The study showed that the D-value of thermo stable PE was 33, 270, 1809, and 1443 s at temperatures 90, 85, 80, and 75°C respectively, and the D-values of thermo labile PE at the corresponding temperatures were 6.3, 22.1, 70.4, and 204.8 s respectively. In addition, the z-values for thermo stable and thermo labile PE determined between 75 and 90°C were found to be 5.47 and 5.75°C, and 9.18 and 9.92°C respectively. A straight line for both thermo stable and thermo labile PE was obtained when D-values were plotted against time-residual activity of PE plotted against T in the range of 75–90°C.

5.9.1 Thermal Inactivation Processes Using Emerging Technologies

5.9.1.1 Case Study 1: High-Pressure Processing as a Pretreatment for Thermosonication

Alicyclobacillus acidoterrestris (AAT) is a rod-shaped, aerobic, gram-positive bacterium. It has the ability to form endospores, which grow at a temperature of 25–60°C and a pH range of 2–7. The optimum growth conditions are a temperature of 40–45°C and a pH range of 4–4.5. AAT spores pose a shelf-life problem for the beverage industry as it has high resistance and can survive thermal pasteurization. As the bacteria do not produce gases during growth, they are difficult to detect. For high-acid fruit products, AAT was recommended as a referral microorganism for the pasteurization process. A study was conducted in which AAT spores were inoculated in orange juice followed by thermosonication (TS). The effects of varying energy density, juice pretreatments, and temperature were investigated. Parameters such as the D-and z-values of spores were calculated for the thermal inactivation processes (Evelyn and Silva, 2016). Experiments were conducted to analyze the effect of TS acoustic energy density (AED) at 75°C on AAT spore inactivation in orange juice. The following steps were undertaken:

- Experiments were conducted to analyze the effect of TS acoustic energy density (AED) at 75°C on AAT spore inactivation in orange juice. Three AEDs, 0.33, 4.10, and 20.20 W/mL, were used and evaluated. The best performance was obtained at 20.20 W/mL.
- TS was conducted at 20.20 W/mL at three temperatures (70, 75, and 78°C) for up to 60 min. The inactivation of AAT was found to be maximum at 78°C.
- The effects of TS on AAT spore inactivation with and without HPP pretreatment of the juice were determined and comparison was made with TS and thermal inactivation at 78°C alone.
- Finally, AAT spores in the orange juice were thermally inactivated at three temperatures (78, 85, and 95°C), following heat shock and 1 min of ultrasound pretreatment.

The results showed that as AED increases, the inactivation of AAT spores also increases. Thus, D_{75} value for 20.20 W/mL was found to be the minimum (49 min). Temperature plays an important role in the inactivation of spores. High temperature produced a marked increase in spore inactivation. It was shown that the D-value was 28 min at 78°C, compared to 49 min and 139 min at 75°C and 70°C respectively at 20.20 W/mL. The study showed that at the same temperature (78°C), the AAT log reduction by TS (2.3 log cycle) was more than with thermal processing (0.3 log cycle). For HPP untreated juice, the D-values at 78°C for TS and thermal processing were found to be 14–28 min and 175 min respectively, thus spore inactivation by TS is more advantageous than by thermal processing at the same temperature. HPP at

600 MPa (4.4 log after 60 min, D78⁰ C-value ¼ 14 min) showed a better performance than that at 200 MPa (2.7 log after 60 min, D78⁰ C-value ¼ 23 min) for the inactivation of spores in orange juice. A further comparison was made between juice subjected to heat shock (HS) followed by 1 min ultrasound, and untreated juice at temperatures 78, 85, and 95°C. Thermal resistance of AAT spores was found to be reduced in case of juice pretreated with HS + ultrasonication. The D-value at 85°C reduced from 69 to 29.3 min and at 78°C it reduced from 175.4 to 103.1 min. TS was found to show better AAT spore inactivation than thermal processing, producing a 6-fold reduction in D-value at same temperature. No marked difference was observed in spore inactivation at TS 70°C and thermal 78°C. Spores were found to be more susceptible to changes in temperature in the thermal process (z-value 8°C) than in TS (z-value 11.5°C). Finally, it was found that to ensure 5D or 6D pasteurization, TS (20.2 W/mL) of at least 2.8 min at 95°C or 23 s at 105°C was recommended to ensure a minimum 3D log reduction on AAT spores in orange juice.

The study demonstrated the advantage of high pressure-assisted thermosonication for the inactivation of *A. acidoterrestris* spores in orange juice. Temperature played an important role in the inactivation of spores. A marked increase in spore inactivation occurred at high temperature. The D-values were shown to be 28 min, 49 min, and 139 min at 78, 75, and 70°C respectively under AED 20.20 W/mL. The results indicated that spore inactivation by TS was more advantageous than by thermal processing at the same temperature. The study showed that both HPP pretreatments improved inactivation of the AAT spores in orange juice and HPP at 600 MPa performed better than at 200 MPa. The thermosonication and thermal spore inactivation followed first-order kinetics. The results demonstrated the advantage of high pressure-assisted thermosonication for the inactivation of *A. acidoterrestris* spores in orange juice.

5.9.1.2 Case Study 2: Application of High Pressure in Thermal Sterilization

In thermal processing, high treatment time and temperature lead to negative effects in the quality of the food products. Emerging technologies such as high-pressure processing are implemented to make food microbiologically safe as well as to avoid undesirable effects on sensory and nutritional properties of food. The target for commercial sterilization in the canning industry is a 12D reduction in the spore population of *Clostridium botulinum.* HPP should therefore show similar spore inactivation or better inactivation behavior. High-pressure thermal pasteurization has been very commonly used for processing of liquid food (Cheftel, 1995; Hogan et al., 2005; Patterson, 2005; Ramaswamy et al., 2010). High-pressure thermal sterilization (HPTS) can also be used for low-acid food production (Aouadhi et al., 2013; Koutchma et al., 2005; Lenz and Vogel, 2014, 2015; Reddy et al., 2006; Sevenich et al., 2013, 2014; Zhu et al., 2008).

Studies on HPP have shown that the spores *of Bacillus amyloliquifaciens* and *Geobacillusste arothermophillus* as well as *Clostridium botulinum, C. sporogenes,* and *C. perfringens* have high resistance to pressure and temperature (Ahn et al., 2007; Ahn et al., 2015; Juliano et al., 2009; Lenz and Vogel, 2014, 2015; Paredes-Sabja et al., 2007; Ramirez et al., 2009; Reineke et al., 2013; Wimalaratne and Farid, 2008). HPTS technology combines the effect of both pressure (400–600 MPa) and temperature (90–121°C) and thus showed a high degree of stability and safety of food products with retention of organoleptic and nutritional properties of food (Knoerzer et al., 2007; Matser et al., 2004; Sevenich et al., 2013; Somerville and Balasubramaniam, 2009).

Food systems have a protective effect for spores by interacting with them (fats, proteins, sugars, salts, etc.) (Olivier et al., 2011). *Geobacillus stearothermophilus* is a rod-shaped, gram-positive bacterium, which causes spoilage in food products and is highly resistant to heat. *G. stearothermophilus* is responsible for the flat sour flavour of canned food products. Due to its high resistance, spores of *Geobacillus stearothermophilus* are used as a biological indicator of thermal sterilization effectiveness (Watanabe et al., 2003). A study was conducted to measure the effectiveness of HPTS for the inactivation of spores of *G. stearothermophilus* in different ready-to-eat (RTE) food products, and the HPTS parameters required to obtain 5D reduction of *G. stearothermophilus* under industrial conditions were calculated. Finally, a study comparing the effects of heat, pressure, and HPTS on spore inactivation in RTE foods was carried out by Ramos et al. (2018). For the study, four different strains of *G. stearothermophilus* were taken: CECT 43 (ATCC 12980), CECT 47, CECT 48, and CECT 4517 (Spanish Type Culture Collection, Valencia, Spain). In addition, two wild strains were extracted from braised veal and tomato soup, and their resistance compared with non-wild strains. Green peas with ham, vegetable cream, steamed sole, and braised veal were selected to represent low-acid foods. In each of the RTE meals, 10 g of 2×10^8 CFU/mL of *G. stearothermophilus* was inoculated followed by packaging in sterile bags which were heat sealed under vacuum. The experiments were conducted using an HPP unit with maximum pressure capacity of 800 MPa and temperature of 130°C with a transmitting medium of propylene glycol/water in the ratio of 70:30 (v/v) and a pressure release time less than 30 min. Preheated samples were loaded into the high-pressure chamber with a pressure of 500–600 MPa and temperature of 110, 115, or 121°C for holding times of 1, 3, or 5 min. Temperatures were recorded during preheating, coming up to pressure, holding, depressurization, and transfer to ice bath.

A 5D reduction of *G. stearothermophilus* spores was taken for a comparative study of HPTS with thermal treatment. In the thermal processing unit, 350 mL of the sample was taken and heated up to stable temperature (110, 115, or 121°C). This was followed by inoculation of 0.2 mL of suspension of microbial cells, then pour plating was carried out with 0.2 mL of sample after different heating times. At constant temperature, temperature inactivation kinetics were studied using the first-order kinetic model.

The study revealed that out of the four strains of *G. stearothermophilus*, CECT 48 had the highest resistance to inactivation when the four RTE foods were subjected to 500 MPa and 110°C, which indicated that CECT 48 strain showed the highest resistance independent of composition of food and time. Similarly, when resistance was checked against 600 MPa pressure conditions at 110°C and 600 MPa and 115°C, the CECT 48 strain was found to be the most resistant. It was observed that when CECT 48 strain spores of *G. stearothermophilus* were subjected to 600 MPa pressure and temperatures of 110, 115, and 121°C, the degree of inactivation of *G. stearothermophilus* during the come-up time varied in different food products. It was found in the experiment that come-up time alone was sufficient to achieve 5D reduction of *G. stearothermophilus* spores at a temperature of 121°C and pressure 600 MPa without using holding time. Thus, it can be suggested that come-up time also acted as an important parameter for the inactivation of spores. It was also reported that during pressure come-up time, different spores showed different resistances. In many studies, compression during pressurization has been shown to contribute to inactivation of microorganisms. Thus, heating during compression cannot be ignored. For comparison purposes, therefore, pressure come-up time and log cycle reduction should be taken into account.

In the study, it was found that there was 3–4D log reduction of *G. stearothermophilus* spores when subjected to HPTS at 110°C and 600 MPa within one minute. The inactivation increased to 5D log reduction when the processing time increased to 5 min. Lethality was found to increase further with increase in temperature to 115°C. It was observed that 5D reduction of *G. stearothermophilus* spores at 115°C and 600 MPa required 1 min in vegetable cream and steamed sole, whereas in peas with ham and braised ham, it required 3 min. Thus, it was reported that depending upon product composition, increase in temperature led to increase in inactivation, hence reduction in processing time.

In addition, it was found that at the same processing temperature, *G. stearothermophilus* CECT 48 spores showed lower resistance to HPTS than to thermal processing. 5D reduction in vegetable cream, peas with ham, braised veal, and steamed sole at 115°C took 53, 37.3, 40.2, and 38.2 min respectively; however, when subjected to HPTS at 600 MPa and 115°C, the 5D value was 1–3 min for the same food systems. At 121°C, the D-values for gaining 5D in vegetable cream, peas with ham, braised veal, and steamed sole were 11.9, 8.85, 11.9, and 10.1 min respectively, but with HPTS at 600 MPa and 121°C, 5D was attained in the first few seconds. The study revealed the synergistic effect of pressure and temperature in HPTS, further shortening the processing time for inactivating spores of *G. stearothermophilus.*

It has also been reported that factors in the composition of food such as fats, salts, proteins, divalent cations, sugars, and water activity retarded inactivation of spores of *G. stearothermophilus* when applied to pressure treatment. As the percentage of fat in the food product increased, the effectiveness in the inactivation of spores of *G. stearothermophilus* decreased. In addition, it was reported that with an increase in processing time, the baroprotective effect of fats decreased. The study showed that in low-acid food having high amounts of fat and proteins, *G. stearothermophilus* spores were inactivated at 115°C, 600 MPa when treated using HPTS for 3 min.

5.9.1.3 Case Study 3: Pressure-Assisted Thermal Stabilization Techniques

A study was carried out to test pressure-assisted thermal stabilization techniques for inactivating spores of *G. stearothermophilus* and *B. cereus* (Wimalaratne and Farid, 2008). In addition, the effectiveness of this technique was studied for vegetative microorganisms of *S. cerevisiae* and *E. coli* present in water. The research was done by comparing D-value (decimal reduction time) for bacteria treated by thermal and by pressure-assisted thermal stabilization (Wimalaratne and Farid, 2008).

Cultures of *G. stearothermophilus*, *B. cereus*, *S. cerevisiae*, and *E. coli* were grown in appropriate media followed by incubation for 10 days at 55°C, 37°C, 30°C, and 37°C respectively. Spores and vegetative cells were separated by centrifugation and then stored at 4°C followed by mixing in water and milk for use during experiment. Inoculated samples were pumped into the apparatus and the temperature increased to 130°C for 3 min. Pressure and temperature were checked every minute. Spore survival was determined based on impedance analysis. From a standard plate count, a calibration curve was made to determine the impedance threshold. Both treated and untreated samples were analyzed.

It was found that the D-values of *G. stearothermophilus* were reduced in pressure-assisted thermal treatment as compared to the thermal process. Similar results were found for *B. cereus,* but to a somewhat lesser extent than for *G. stearothermophilus*. As *S. cerevisiae* and *E. coli* do not undergo the two inactivation steps, no effect of pressure and heat was observed on them (Wimalaratne and Farid, 2008). It was observed that the z-values of *G. stearothermophilus* in Milli-Q water treated thermally and by pressure-assisted thermal sterilization were 8.4°C and 10.4°C respectively. Similarly, in milk z-values were found to be 5.8 and 6.3°C for thermal treatment and pressure-assisted thermal sterilization respectively. The z-values for *B. cereus* in Milli-Q water were found to be 12.5 and 8.3°C. It was observed that pressure-assisted thermal sterilization decreased D-values.

5.9.1.4 Case Study 4: Application of High Pressure to Inactivate Polyphenol Oxidase

Browning in fruits and vegetables mainly occurs because of an enzyme called PPO (polyphenol oxidase). PPO causes hydroxylation of monophenols, which leads to formation of o-diphenols, which subsequently undergo oxidation to o-quinines, which finally undergo polymerization to form brown pigments (Espín et al., 1998). This enzymatic browning causes change in color, which ultimately affects customer purchasing decisions. PPO also degrades polyphenols, which causes deterioration of food quality. PPO therefore needs to be inactivated to maintain food quality. PPO is an enzyme containing copper and it exists in two forms: membrane bound and soluble (Liu et al., 2015; Zaini et al., 2013). In food industries, thermal processing above 80°C is used to inactivate PPO (Gouzi et al., 2012; Zhou et al., 2018; Tomas-Barberan and Espín, 2001).

Studies were conducted on the inactivation kinetics of soluble and membrane-bound PPO in pear using thermal and HPP techniques (Zhou et al., 2018). Both soluble and membrane-bound PPO showed maximum activity at pH of 6.0 and 6.5 respectively. It was seen that when pH increased above 7, there was a decrease in the activity of both PPOs; their activity also varied with different varieties of pear (Zhou et al., 2018). Thermostability was determined for both soluble and membrane-bound PPO when they were subjected to temperaturesof 25–100°C. At pH 4.5, soluble and membrane-bound PPO showed stable activity from 25–50°C but as the temperature increased above 60°C, the activity was drastically reduced. At a temperature above 80°C, PPO was completely inactivated after 10 min treatment time. Similarly, when membrane-bound PPO were subjected to temperature 40–70°C for 10 min, the residual activity of membrane-bound PPO was found to be lower than that of soluble PPO at 4.4 pH, which showed that membrane-bound PPO had lower thermal stability than soluble PPO. At lower pH, i.e., 3.5, PPO activity was affected as at this pH they become more sensitive to temperature. However, in phosphate buffer (pH 6.5), it was seen that enzyme activity decreased at a temperature higher than 40°C, and enzymes were inactivated when treated at 70–80°C for 10 min (UlkerYerliturk et al., 2008). This reduction in thermostability of PPO at acidic pH was due to the structure of PPO disintegrating. From the study, it was observed that D-value was 187.5 min for soluble PPO at pH 4.4 at a temperature 60°C. There was a reduction in D-value for soluble PPO as temperature was increased from 60 to 75°C. Similar results were obtained for membrane-bound PPO at pH 4.4. It was found that membrane-bound PPO has much lower D-value than that of soluble PPO, which showed that membrane-bound PPO of pear was more susceptible to thermal inactivation than soluble PPO regardless of pH. In addition, it was observed that at pH 4.4 and 3.5, soluble and membrane-bound PPO have z-values of 10.1°C and 12.5°C respectively; at 3.5 pH, it was 15.2°C and 13.5°C respectively. A low z-value showed higher sensitivity of enzymes to temperature change at lower pH (Gouzi et al., 2012; Weemaes et al., 1998).

High-pressure processing was conducted at 600 MPa and at a temperature range of 25–80°C. HPP at 25°C was found to cause much less inactivation of PPO. At pH 4.4, soluble PPO were resistant to pressure but membrane-bound PPO showed higher sensitivity towards inactivation at 25°C. At higher temperature, inactivation of membrane-bound PPO was observed. The study also showed that inactivation of soluble PPO at pH 3.5 was the same as that of membrane-bound PPO at pH 4.4, indicating the pressure-resistant nature of soluble PPO regardless of pH. But when thermal treatment at 40, 50 and 60°C was combined with high pressure, it was seen that at pH 3.5, soluble PPO was inactivated. Thus, when a combination of HPP and thermal treatment was applied at 60°C for 10 min, complete inactivation of soluble PPO was observed at pH 3.5. In the study, z-values for soluble and membrane-bound PPO against high-pressure thermal processing at pH 4.4 and pH 3.5 were found to be 31.1, 41.6, 16.1, and 26.9°C respectively. These values were greater than the z-values obtained during thermal processing, indicating reduced sensitivity to high temperature when subjected to high-pressure thermal processing. A synergistic effect of high pressure and temperature on the inactivation of enzymeswas observed at higher pressure (>400 MPa).

5.9.1.5 Case Study 5: Surface Pasteurization Treatment for Thermal Inactivation

Listeriosis monocytogenes is responsible for contamination of poultry products. The United States Department of Agriculture (USDA) has made zero tolerance for *L. monocytogenes* compulsory in RTE meat products. In order to develop an effective in-package thermal treatment process for meat products, D-values and z-values of targeted microorganisms (*L. mononocytogenes* and *Salmonella typhimurium*) were evaluated at various stages of processing. To ensure delivery of safe poultry products to consumers, a study was carried out to determine the surface thermal inactivation kinetic values of *L. monocytogenes* and *Salmonella typhimurium* on packaged turkey bologna (McCormick et al., 2003).

Samples of meat were inoculated with 0.1 mL of either *L. monocytogenes* or *S. typhimurium* followed by vacuum sealing in low-density polyethylene bags. Heating data including time and temperature were measured. Due to low recoverability of cells at a higher processing temperature, 61 and 65°C were selected for *L. monocytogenes* and 57 and 60°C for *S. typhimurium*. A thermal death curve was plotted. From the research, it was found that average come-up times were approximately 60 s for all water-bath temperatures. At 65°C of water bath temperature, the lethality was found to be 1 log, whereas at other temperatures lethality was found to be 0.5 log. For *L. monocytogenes*, the D-values at 61 and 65°C were found to be 124 s and 16.2 s respectively with a z-value of 4.4°C. It was observed that when packaged meat was submerged for 10 s in 85°C water, no *L. monocytogenes* cells were detected, but more than 100 cells were found if the meat was submerged at 61°C for 10 min.

A study showed that thermal resistance of bacteria is affected by fat content. D-values of low-fat food products are 8–10 times lower than those of high-fat foods (Doyle et al., 2001). Thermal resistance was also affected by the phase of the bacterial growth curve. Cells present in the exponential phase are less likely to be resistant thermally than cells in the stationary phase (Doyle et al., 2001). It was found in the study that D-values for *S. typhimurium* at 57°C and 60°C were 278 and 81 s respectively, and the z-values were 5.6°C. It can therefore be concluded that thermal resistance (D-values) of *S. typhimurium* and *L. monocytogenes* in RTE meat products is 124 s and 16.2 s at 61°C and 65°C respectively for *L. monocytogenes* and 278 s and 81 s at 57 and 60°C respectively for *S. typhimurium*. When calculating D-values, factors such as food substrate, growth conditions, growth phase, and strain should be taken into account (McCormick et al., 2003).

5.10 CONCLUDING REMARKS AND FUTURE SCOPE

The D-value, z-value and F-value are the biological indicators that must be configured to meet requirements concerning survival of microbial spores in processed foods. Safety and quality of the food during food processing and preservation must be considered. Different food technologies such as drying, pasteurization, and sterilization can ensure microbiological safety or stability of the products. Non-thermal technologies such as high hydrostatic pressure, pulsed electrical fields, ultrasound, and UV light can achieve the desired quality at shorter processing times and lower temperature levels, so these processes can ensure the nutritional value as well as extended shelf life of food products. Development of appropriate models to describe survivor curves for microbial populations using different thermal and non-thermal preservation technologies would be beneficial. These models would establish the synergistic effects of high hydrostatic pressure, pulsed electrical fields, ultrasound, and UV light with temperature for inactivation of microorganisms.

REFERENCES

Ahn, J., Balasubramaniam, V., and Yousef, A. 2007. Inactivation kinetics of selected aerobic and anaerobic bacterial spores by pressure-assisted thermal processing. *International Journal of Food Mmicrobiology*, 113(3): 321–329.

Ahn, J., Lee, H., and Balasubramaniam, V. 2015. Inactivation of *Geobacillusstearothermophilus* spores in low-acid foods by pressure-assisted thermal processing. *Journal of the Science of Food and Agriculture*, 95(1): 174–178.

Aouadhi, C., Simonin, H., Prévost, H., de Lamballerie, M., Maaroufi, A., and Mejri, S. 2013. Inactivation of *Bacillus sporothermodurans* LTIS27 spores by high hydrostatic pressure and moderate heat studied by response surface methodology. *LWT-Food Science and Technology*, 50(1): 50–56.

Atkins, C.D. and Rouse, A.H. 1953. Time–temperature relationships for heat inactivation of pectinesterases in citrus juices. *Food Technology*, 7: 489–491.

Bahceci, K.S. and Acar, J. 2007. Determination of guaiacol produced by Alicyclobacillusacidoterrestris in apple juice by using HPLC and spectrophotometric methods, and mathematical modeling of guaiacol production. *European Food Research and Technology*, 225(5–6): 873–878.

Barrett, D.M. and Gonzalez, C. 1994. Activity of softening enzymes during cherry maturation. *Journal of Food Science*, 59: 574–577.

Cheftel, J.C. 1995. High-pressure, microbial inactivation and food preservation. *Revista de Agaroquimica y Tecnologia de Alimentos*, 1(2–3): 75–90.

Chen, C.S., Shaw, P.E., and Parish, M.E. 1993. Orange and tangerine juices. In: Nagy, S., Chen, C.S. and Shaw, P.Z. (eds.), *Fruit Juice Processing Technology*. Aubumdale, FL: Agscience Inc., pp. 119–124.

De Carvalho, A.A.T., Vanetti, M.C.D., and Mantovani, H.C. 2008. Bovicin HC5 reduces thermal resistance of Alicyclobacillusacidoterrestris in acidic mango pulp. *Journal of Applied Microbiology*, 104(6): 1685–1691.

Deak, T. 2014. Food Technology: Sterilization. In: *Encyclopedia of Food Safety*, Vol. 3. Amsterdam: Elsevier Inc., pp. 245–252.

Doyle, M.E., A.S. Mazzotta, T. Want, D.M. Wiseman, and V.N. Scott. 2001. Heat resistance of *Listeria monocytogenes*. *Journal of Food Protection*, 64: 419–429.

Eagerman, B.A. and Rouse, A.H. 1976. Heat inactivation temperature– Time relationships for pectinesterase inactivation in citrus juices. *Journal of Food Science*, 41: 1396–1399.

Espín, J.C., Jolivet, S., and Wichers, H.J. 1998. Inhibition of mushroom polyphenol oxidase by agaritine. *Journal of Agricultural and Food Chemistry*, 46(8): 2976–2980.

Evelyn, E., and Silva, F.V.M. 2016. High pressure processing pretreatment enhanced the thermosonication inactivation of *Alicyclobacillusacidoterrestris* spores in orange juice. *Food Control*, 62: 365–372.

Ghani, A.A., Farid, M., and Chen, X. 2002. Theoretical and experimental investigation of the thermal destruction of Vitamin C in food pouches. *Computers and Electronics in Agriculture*, 34(1): 129–143.

Gould, G.W., and Sale, A.J.H. 1970. Initiation of germination of bacterial spores by hydrostatic pressure. *Journal of General Microbiology*, 60: 335.

Gouzi, H., Depagne, C., and Coradin, T. 2012. Kinetics and thermodynamics of the thermal inactivation of polyphenol oxidase in an aqueous extract from *Agaricusbisporus*. *Journal of Agricultural and Food Chemistry*, 60(1): 500–506.

Hogan, E., Kelly, A.L., Da-Wen, S. 2005. High pressure processing of foods: an overview. In: Da-Wen, S (ed.), *Emerging Technologies for Food Processing*. Amsterdam: Elsevier Science, pp. 3–32.

Juliano, P., Knoerzer, K., Fryer, P.J., and Versteeg, C. 2009. *C. botulinum* inactivation kinetics implemented in a computational model of a high-pressure sterilization process. *Biotechnology Progress*, 25(1): 163–175.

Juneja, V.K., Snyder, O.P., and Marmer, B.S. 1997. Thermal destruction of *Escherichia coli 0157:H7* in beef and chicken: determination of D- and z-values. *International Journal of Food Microbiology*, 35: 231–237.

Khue, D.N., Tiep, H.T., and Dat, L.Q. 2020. Influence of frequency and temperature on the inactivation of Salmonella entericaserovarenteritidis in Ohmic heating of pomelo juice. *LWT*, 109528.

Knoerzer, K., Juliano, P., Gladman, S., Versteeg, C., and Fryer, P.J. 2007. A computational model for temperature and sterility distributions in a pilot-scale high-pressure high-temperature process. *AIChE Journal*, 53(11): 2996–3010.

Komitopoulou, E., Boziaris, I.S., Davies, E.A., Delves-Broughton, J., and Adams, M.R. 1999. Alicyclobacillusacidoterrestris in fruit juices and its control by nisin. *International Journal of Food Science &Technology*, 34(1): 81–85.

Koutchma, T., Guo, B., Patazca, E. Parisi, B. 2005. High pressure–high temperature sterilization: from kinetic analysis to process verification. *Journal of Food Process Engineering*, 28(6): 610–629.

Lee, J.Y., Lin, Y.S., Chang, H.N., Chen, W., and Wu, M.C., 2003. Temperature–time relationships for thermal inactivation of pectinesterases in orange juice. *Journal of Science and Food and Agriculture*, 83: 681–684.

Lenz, C.A. and Vogel, R.F. 2014. Effect of sporulation medium and its divalent cation content on the heat and high pressure resistance of *Clostridium botulinum* type E spores. *Food Microbiology*, 44: 156–167.

Lenz, C.A. and Vogel, R.F. 2015. Differential effects of sporulation temperature on the high pressure resistance of *Clostridium botulinum* type E spores and the interconnection with sporulation medium cation contents. *Food Microbiology*, 46: 434–442.

Lin, S.H. 1976. Continuous high-temperature/short-time sterilization of liquid foods with steam-injection heating. *Chemical Engineering Science*, 31(1): 77–82.

Liu, F., Zhao, J.H., Gan, Z.L., and Ni, Y.Y. 2015. Comparison of membrane-bound and soluble polyphenol oxidase in Fuji apple (MalusdomesticaBorkh. cv. Red Fuji). *Food Chemistry*, 173: 86–91.

Lopez, M.D., García, P., Munoz-Cuevas, M., Fernández, P.S., and Palop, A. 2011. Thermal inactivation of Alicyclobacillusacidoterrestris spores under conditions simulating industrial heating processes of tangerine vesicles and its use in time temperature integrators. *European Food Research and Technology*, 232(5): 821–827.

Matser, A.M., Krebbers, B., van den Berg, R.W., and Bartels, P.V. 2004. Advantages of high pressure sterilization on quality of food products. *Trends in Food Science & Technology*, 15(2): 79–85.

Mazzotta, A.S. 2001. Thermal inactivation of stationary-phase and acid-adapted Escherichia coli O157:H7, Salmonella, and Listeria monocytogenes in fruit juices. *Journal of Food Protection*, *64*(3): 315–320.

McCormick, K., Han, I.Y., Acton, J.C., Sheldon, B.W., and Dawson, P.L. 2003. *D*- and *z*-Values for *Listeria monocytogenes* and *Salmonella typhimurium* in packaged low-fat ready-to-eat turkey bologna subjected to a surface pasteurization treatment. *Poultical Science*, 82(8): 1337–1342.

McKnight, I.C., Eiroa, M.N.U., Sant'Ana, A.S., and Massaguer, P.R. (2010). Alicyclobacillusacidoterrestris in pasteurized exotic Brazilian fruit juices: Isolation, genotypic characterization and heat resistance. *Food Microbiology*, 27(8): 1016–1022.

Nayak, B., Berrios, J.D.J., Powers, J.R., and Tang, J.M. 2011. Thermal degradation of anthocyanins from purple potato (cv. Purple Majesty) and impact on antioxidant capacity. *Journal of Agricultural and Food Chemistry*, 59(20): 11040–11049.

Olivier, S.A., Bull, M.K., Stone, G., van Diepenbeek, R.J., Kormelink, F., Jacops, L., and Chapman, B. 2011. Strong and consistently synergistic inactivation of spores of spoilage-associated *Bacillus* and *Geobacillus spp.* by high pressure and heat compared with inactivation by heat alone. *Applied and Environmental Microbiology*, 77(7): 2317–2324.

Parades, S.D., Gonzalez, S., Sarkar, M.R., and Torres, J.A. 2007. Combined effects of hydrostatic pressure, temperature, and ph on the inactivation of spores of *Clostridium perfringens* Type A and *Clostridium sporogenes* in buffer solutions. *Journal of Food Science*, 72(6): M202–M206.

Patterson, M.F. 2005. Microbiology of pressure-treated foods. *Journal of Applied Microbiology*, 98(6): 1400–1409.

Ramaswamy, H.S., Shao, Y., and Zhu, S. 2010. High-pressure destruction kinetics of *Clostridium sporogenes* ATCC 11437 spores in milk at elevated quasi-isothermal conditions. *Journal of Food Engineering*, 96(2): 249–257.

Ramirez, R., Saraiva, J., Lamela, C.P., and Torres, J.A. 2009. Reaction kinetics analysis of chemical changes in pressure-assisted thermal processing. *Food Engineering Reviews*, 1(1): 16–30.

Ramos, S.J., Millan, D., Ortiz, L., Alonso, D., and Garcia de la Torre, S. 2018. Effect of High-Pressure Thermal Sterilization on the Inactivation of *Geobacillusstearothermophilus* spores in ready to eat meals. *The International Journal of Engineering and Science:* 65–74.

Reddy, N., Tetzloff, R., Solomon, H., and Larkin, J. 2006. Inactivation of *Clostridium botulinum* non proteolytic type B spores by high pressure processing at moderate to elevated high temperatures. *Innovative Food Science &Emerging Technologies*, 7(3): 169–175.

Reineke, K., Mathys, A., Heinz, V., and Knorr, D. 2013. Mechanisms of endospore inactivation under high pressure. *Trends in Microbiology*, 21(6): 296–304.

Sevenich, R., Bark, F., Crews, C., Anderson, W., Pye, C., Riddellova, K., Hradecky, J., Moravcova, E., Reineke, K., and Knorr, D. 2013. Effect of high pressure thermal sterilization on the formation of food processing contaminants. *Innovative Food Science & Emerging Technologies*, 20: 42–50.

Sevenich, R., Kleinstueck, E., Crews, C., Anderson, W., Pye, C., Riddellova, K., Hradecky, J., Moravcova, E., Reineke, K., and Knorr, D. 2014. High-pressure thermal sterilization: food safety and food quality of baby food puree. *Journal of Food Science*, 79(2): M230–M237.

Shao, Y. and Ramaswamy, H.S. 2011. *Clostridium sporogenes*-ATCC 7955 spore destruction kinetics in milk under high pressure and elevated temperature treatment conditions. *Food and Bioprocess Technology*, 4(3): 458–468.

Silva, F.M., Gibbs, P., Vieira, M.C., and Silva, C.L. 1999. Thermal inactivation of Alicyclobacillusacidoterrestris spores under different temperature, soluble solids and pH conditions for the design of fruit processes. *International Journal of Food Microbiology*, 51(2–3): 95–103.

Silva, F.V., Tan, E.K., and Farid, M. 2012. Bacterial spore inactivation at 45–65° C using high pressure processing: Study of Alicyclobacillus acidoterrestris in orange juice. *Food Microbiology*, 32(1): 206–211.

Silva, F.V.M. and Gibbs, P.A. 2001. *Alicyclobacillusacidoterrestris* spores in fruit products and design of pasteurization processes. *Trends in Food Science & Technology*, 12(2): 68 e74.

Silva, F.V.M. and Gibbs, P.A. 2010. Non-proteolytic *Clostridium Botulinum* spores in low-acid cold-distributed foods and design of pasteurization processes. *Trends in Food Science and Technology*, 21: 95–105.

Somerville, J. and Balasubramaniam, V.B. 2009. Food safety series – Pressure-assisted thermal sterilization of low-acid, shelf-stable foods. *Resource Magazine*, 16(7).

Splittstoesser, D.F., Churey, J.J., and Lee, C.Y. 1994. Growth characteristics of aciduricspore-forming bacilli isolated from fruit juices. *Journal of Food Protection*, 57(12): 1080–1083.

Sulaiman, A., Soo, M.J., Farid, M., and Silva, F.V. 2015. Thermosonication for polyphenoloxidase inactivation in fruits: Modeling the ultrasound and thermal kinetics in pear, apple and strawberry purees at different temperatures. *Journal of Food Engineering*, 165: 133–140.

Tarr, P.I. 1994. Review of 1993 *Edwrichia coli* 0157:H7 outbreak: Western US. *Dairy Food Environmental and Sanitation*, 14: 372–373.

Tomas-Barberan, F.A. and Espín, J.C. 2001. Phenolic compounds and related enzymes as determinants of quality in fruits and vegetables. *Journal of the Science of Food and Agriculture*, 81(9): 853–876.

UlkerYerliturk, F., Arslan, O., Sinan, S., Gencer, N., and Ozensoy, G. 2008. Characterization of polyphenoloxidase from wild pear (*Pyruselaegnifolia*). *Journal of Food Biochemistry*, 32(3): 368–383.

Verbeyst, L., Van Crombruggen, K., Van der Plancken, I., Hendrickx, M., and Van Loey, A. 2011. Anthocyanin degradation kinetics during thermal and high pressure treatments of raspberries. *Journal of Food Engineering*, 105(3): 513–521.

Vikram, V.B., Ramesh, M.N., and Prapulla, S.G. 2005. Thermal degradation kinetics of nutrients in orange juice heated by electromagnetic and conventional methods. *Journal of Food Engineering*, 69(1): 31–40.

Vishwasrao, C. and Ananthanarayan, L. 2018. Kinetics of inactivation of quality-deteriorating enzymes and degradation of selective phytoconstituents in pink guava pulp during thermal processing. *Journal of Food Science and Technology*, 55(8): 3273–3280.

Walls, I. 1997. *Alicyclobacillus-an overview*. In: *Institute of Food Technologists Annual Meeting in Orlando*, Session 36-1, Florida, June 14–18.

Watanabe, T., Furukawa, S., Hirata, J., Koyama, T., Ogihara, H., and Yamasaki, M. 2003. Inactivation of *Geobacillusstearothermophilus*spores by high-pressure carbon dioxide treatment. *Applied and Environmental Microbiology*, 69(12): 7124–7129.

Weemaes, C., Ludikhuyze, L., Van den Broeck, I., and Hendrickx, M.E. 1998. Effect of pH on pressure and thermal inactivation of polyphenol oxidases: A kinetic study. *Journal of Agricultural and Food Chemistry*, 46(7): 2785–2792.

Wimalaratne, S. and Farid, M. 2008. Pressure assisted thermal sterilization. *Food and Bioproducts Processing*, 86(4): 312–316.

Zaini, N.A., Osman, A., Hamid, A.A., Ebrahimpour, A., and Saari, N. 2013. Purification and characterization of membrane-bound polyphenoloxidase (mPPO) from snake fruit [Salaccazalacca (Gaertn.) Voss]. *Food Chemistry*, 136(2): 407–414.

Zhou, L., Liu, W., and Terefe, N.S. 2018. The inactivation kinetics of soluble and membrane-bound polyphenol oxidase in pear during thermal and high-pressure processing. *Food and Bioprocess Technology*, 11(5): 1039–1049.

Zhu, S., Naim, F., Marcotte, M., Ramaswamy, H., and Shao, Y. 2008. High-pressure destruction kinetics of *Clostridium sporogenes* spores in ground beef at elevated temperatures. *International Journal of Food Microbiology*, 126(1): 86–92.

6 Fundamentals and Applications of Deep-Fat Frying

Ajita Tiwari
Department of Agricultural Engineering, Triguna Sen School of Technology, Assam University, Silchar, Assam, India
Puja Das
Department of Agricultural Engineering, Triguna Sen School of Technology, Assam University, Silchar, Assam, India

CONTENTS

6.1 INTRODUCTION

Frying is among the most primitive yet nevertheless most efficient and effective food preparation techniques. It is popular for the unique flavor, sensory quality, and taste it gives to foods (Ngadi and Xue, 2009). Changes in sensory and nutritional qualities are caused by simultaneous heat and mass transfer processes resulting in the complex interaction of food and oil (Ziaiifar et al., 2008). Its preservative action is caused by high levels of heat generation which cause inactivation of microorganisms and enzymes, and rapid loss of moisture content on the surface of the food (Fellows, 2006). Application of improved frying technologies is helping to reduce deterioration in quality and nutrition; some engineering designs and considerations also help to improve cooking quality during deep-fat frying.

Deep-fat frying (DFF) is a multifunctional operation defined as cooking food by immersion in high-temperature (165–190°C) edible oil or fat above the boiling point of water (Hubbard and Farkas, 2000). It can be a good alternative as a drying operation because of its rapid simultaneous heat and mass transfer. In general, as well as food applications, this process is used to dry and stabilize meat carcasses, sewage sludge, and wood. The DFF operation is a complex one (Figure 6.1). In this process, two mass transfers in opposite directions are taking place within the material being fried: the water and some soluble material escape from the starchy products and the oil enters the food (Blumenthal, 1991).

There is a significant fat content in meat or fish but during the process of raw material frying in an oil bath, sometimes even fat can escape from the product (Oroszvari et al., 2005). Moreover, the material experiences chemical and physical changes because of the high operating temperature range of frying (140–180°C). Frying links dehydration to cooking through starch gelatinization, protein denaturation,

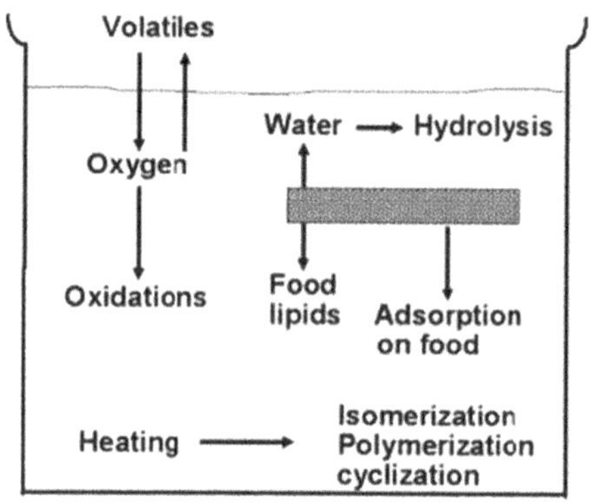

FIGURE 6.1 Major reactions during deep frying

aromatizing, changes in color/appearance via Maillard reactions, and oil uptake. These reactions are strongly inter-related but can be transformed by adaptation and optimization of process parameters (Vitrac, 2000). Currently, the major critical nutritional point of DFF is fat uptake. Therefore, there is a need to improve the frying process by either controlling the final fat content of the fried product or minimizing it. In addition, during frying in the oil bath, oil deteriorates due to the simultaneous presence of oxygen, heat, and water, and consequently potential newly formed toxic compounds may appear (Gertz, 2005). These newly formed compounds alter the physical and chemical properties of oil and can enhance the oil uptake while frying. Thus, the quality and quantity of oil uptake requires careful control.

6.2 FRYING MECHANISM

6.2.1 Deep-Fat Frying or Immersion Frying

DFF was first developed due to the influence of olive oil around the Mediterranean region, but nowadays many processed foods are deep-fat fried due to the unique flavor and texture imparted to the food by the process. It is also known as immersion frying (Varela, 1988). A large amount of fried foods are used at industrial and commercial scale and are very popular with consumers. Oil and fat consumption are high largely because DFF food products, crucially, absorb a high amount of oil during the process, contributing 40% of total weight of the fried product.

To optimize the health benefits of their diet and eliminate trans-fat, consumers should focus on consuming food products that are low in sodium, fat, calories, and carbohydrates, and are fully organic. Mariscal and Bouchon (2008) state that consumers like fried food because of its intense flavor and unique taste. Most of them do not compromise on taste, hence the important fried food trend in the salty snack market.

6.2.2 Deep-Fat Frying Process

Deep-fat frying is simply the process of cooking food immersed in high-temperature (130–190°C) edible oil, which may be classified as a dehydration process (Farkas, 1994). It involves high temperature, and significant changes in the micro-structural surface and body of the food. At the same time, continuous heat and mass transfer occur in the opposite direction to water loss (Bouchon et al., 2003).

During DFF, water continuously evaporates from the outer layer of food products and into the surrounding oil, causing surface drying which leads to formation of a crust. Before immersion in the oil, food material may be covered with an edible surface layer for good flavor retention. Most of the desirable characteristics of fried food are derived from its composite structure: a dry, porous, crisp, and oily outer layer or crust, and a moist cooked interior (Bouchon and Aguilera, 2001).

Frying technology is important for many sectors of the food industry: suppliers of oils and ingredients, fast-food and restaurant operators, industrial producers of fully fried, part-fried, and snack foods, and manufacturers of frying equipment (Blumenthal, 1991). Deep-fat fryers basically consist of a chamber into which heated oil and food are placed. Their size depends on their use. Frying equipment is divided into two broad categories: (1) batch-frying equipment, normally used in catering and in small plants, and (2) continuous fryers, which are used on an industrial scale to process large amounts of food. Generally, frying operations are carried out at normal atmospheric conditions, but for efficient frying pressure above or below atmospheric may sometimes be used.

6.2.3 Heat and Mass Transfer During Deep-Fat Frying

Simultaneous heat and mass transfer occur during DFF. Heat is transferred from the frying medium to the food surface by convection, and afterwards heat transfer takes place within the food by conduction.

After initial heating, evaporation of water at the surface of the food begins, and the interstitial liquid boiling point, which is higher than the normal boiling point of water, is reached. At this stage the food loses its surface water; the rate of water escape increases rapidly, and due to the high turbulence associated with nucleate boiling, heat transfer through natural convection gives way to forced convection (Figure 6.2). The evaporation front moves towards the interior of the food as frying time increases, causing formation of a dehydrated crust layer. The values around the boiling point of the liquid are restricted due to the temperature of the liquid water present inside the food material. The rate of water loss decreases (falling-rate period) as the frying proceeds, leading to the bubble-end point, which is indicated by the termination of water removal from the food (Singh, 1995). These process conditions are suitable for products like chips, where the permitted moisture content is 5%.

Several mathematical models have been developed to understand the numerous complexities involved in this process. These models mainly deal with frying products by understanding their physical properties. There are numerous models dealing with

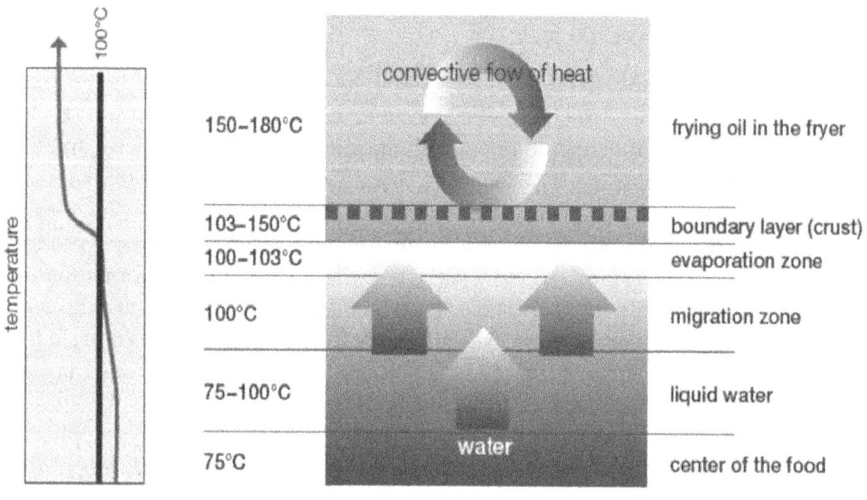

FIGURE 6.2 Heat and mass transfer mechanism of deep-fat frying

the phenomena of heat and mass transfer and various formulae that account for evaporation, either completely or without the transport of oil phase.

Single models (frying) are of two types: 1) semi-empirical (mass and heat balance) models, and 2) differential Equation (DE) models. Semi-empirical modes are treated partially; DE models are treated more fully as they are applicable to frying problems.

6.3 MATHEMATICAL MODELS

6.3.1 Mass and Heat Balance Models (Semi-Empirical Models)

These kinds of models are ordinary DE models that deal with partially treated products where a crust is formed. Freezing is mainly used to determine the thickness of the crust formed. However, these models have a number of limitations: (1) assumed constant value for the physical and thermal properties; (2) oil absorption is not considered; (3) sensible heating in the crust and core is not considered; and (4) moisture profiles and the transient temperature in the core of food are not addressed.

6.3.2 Deterministic Models (De Models)

Models based on the phenomena of heat and mass transfer during frying are known as DE models. The calculations developed by this kind of model can only be solved numerically. The model states that during frying, the initial heat transfer mode is convection, when the oil reaches the product surface; later on, when water moves out from the inner core region, oil is absorbed into the food. During frying water leaves the food product, usually in the form of water vapor, during frying, causing rapid decrease in moisture content and increase in oil content of the food. During frying two regions are formed, the crust and the core, their interface moving into the food throughout the process. For a short time the temperature remains at boiling point in the interface to allow the water present in the food to evaporate.

The assumptions made for these models are 1) initially the product is isothermal and isotropic, i.e., at the initial stage of frying, moisture and temperature are uniformly distributed in the product; 2) the heat requirement is much lower for chemical reaction than for the evaporation of water; 3) physical and thermal properties are dependent parameters of moisture and temperature content during DFF; 4) after frying a microscopically uniform porous medium is formed; 5) after frying a uniform layer of oil is formed which covers the surface of the product; and 6) during cooling, a lot of oil diffuses into fried products.

6.3.3 Mechanistic Models (Multiphase Porous Media Models)

This model, developed on the principles of mass, momentum, and energy conservation, deals with mass transport, kinetics, and thermodynamic equilibrium. The model produces complex equations which can be solved numerically. Some simple assumptions are made to simplify homogeneity, geometry, and boundary conditions. Mechanistic models are mainly based on physics and empirical equation models.

6.4 STRUCTURAL DEVELOPMENT AND CHANGES DURING DFF

Structural development and changes take place during frying because of alterations in the cellular and sub-cellular regions of the crust and core of fried food (Aguilera and Gloria, 1997). After fast frying and cryo-sectioning, fast freezing of food takes place. The finished fried products consist of three distinct microstructures:

(1) The cell wall remnant (damaged cell) forms a thin outer layer approximately 250 μm thin;
(2) The evaporation front is extended due to the intermediate layer formed by shrunken intact cells;
(3) Gelatinized starch is present in the fully hydrated intact cells of the core.

At the same time micro-structural changes occur in the core region during frying. The inner structure restricts the entry temperature to below water boiling point. This process leads to changes in structural properties: protein denaturation, softening of the middle lamellae, and gelatinization of starch.

Due to exposure of the crust (the outer surface of the fried product), the micro-structural changes taking place are more aggressive than those occurring in the inner structure (in atmospheric deep-fat frying). Apart from the physical damage caused by cutting, other chemical and physical changes also occur, such as protein denaturation, water evaporation, starch gelatinization and rapid dehydration of cells located in the forming crust, breakdown of cellular adhesion, subsequent dehydration, and oil uptake (Bouchon and Aguilera, 2001).

Baumann and Escher (1995) explain the oil absorption mechanism and the importance of micro-structural changes occurring during deep-fat frying. The factors affecting the uptake of oil into the food need to be validated through analysis of the structure, including its relation to the area where oil is deposited and also the oil mechanism to the surface of product. The oil uptake mainly involves loss of moisture content, and important aspects of the crust region (micro-structure) include permeability, pore size, and binding properties (Bouchon et al., 2001).

6.5 PROPERTIES OF FRIED FOOD

In food products, fats and oils play an important functional and sensory role. During the frying process fats and oil interact with the other compounds, which improves the quality of products by releasing good flavor, aroma, and taste, and develops unique textural characteristics (Giese, 1996). The main objective of DFF is to make food more acceptable to the consumer. Fat is the best natural agent to enhance flavor and texture. During frying fat enters the food and replaces part of the water, resulting in a considerably more pleasant taste. The outer crust of the food is tenderized and wetted due to absorption of fat. The process contributes to sensory attributes such as crispness (Stevenson et al., 1984).

Lipid degradation products originating from the frying oil gives different fried flavors. Fried products that are golden brown in color can be regarded as properly cooked at optimum temperature and time. Foods that are under-fried, cooked at low temperature and for a short time do not form a proper brown color and have ungelatinized starch content.

6.6 CHEMICAL REACTIONS OF OIL DURING DEEP-FAT FRYING

6.6.1 Oil Hydrolysis

Moisture content in the food product gradually subsides when it is being fried in oil. Initially it forms steam and evaporates with a bubbling action. A chemical reaction is initiated in the DFF process with the interaction of oil and food, involving water, steam, and oxygen. The generation of free fatty acids, diacylglycerols, monoacylglycerols, and glycerol takes place due to the action of water, which is a weak nucleophile, on the ester linkage of triacylglycerols. Fresh oil has less than 1% free fatty acid (FFA) but in repeated frying operations the FFA percentage of oil increases (Chung et al., 2004). The importance of FFA lies mainly in monitoring frying oil quality. Oil hydrolysis is greater with short and unsaturated fatty acids, which are more soluble in water than long and saturated fatty acids.

6.6.2 Oil Oxidation

Oxygen reacts with oil in the DFF process (Cuesta et al., 1993; Peers and Swoboda, 1982; Sanchez-Muniz et al., 1993). The chemical mechanism of thermal oxidation is similar to that of autooxidation, but the rate is much higher.

6.6.3 Oil Polymerization

Volatile compounds at ppm concentrations are important to the flavor of fried food products and frying oil, but they can be totally decomposed during frying (Nawar, 1985). The decomposition products in frying oil are triacylglycerol dimers, nonvolatile polar compounds, and polymers. There are fewer cyclic compounds than dimers, polymers, and nonvolatile polar compounds. Depending on the reaction process and the kinds of fatty acids present in the oil, dimers and polymers may be either cyclic or acyclic (Cuesta et al., 1993). Polymerization and dimerization in DFF result from radical reaction mechanisms.

6.6.4 Effect on the Lipid Fraction of Foods

In some food products, oxidation of lipids takes place during DFF, due to the rapid increase in temperature and oxygen concentration. It is quite difficult to identify the effect on nutritional characteristics, due to the interference of a number of factors, especially the type of oil, its thermal treatment history, and the proportion that is absorbed in the food (Fellows, 2006). Frying should be chosen based on a procedure that retains a good number of nutritional characteristics due to its oil uptake (Mellema, 2003).

High calorific value is one of the problems associated with fried food consumption. Due to retention and absorption of oil, the fat content increases, and this means an increase in food calorific value on average of 53% in hamburgers and 42% in French fries (USDA, 2013). If food products are repeatedly fried in reused oils, the fat digestibility keeps on changing, with adverse impacts on health. Toxic compounds may also be generated and found in oil, even if a frying oil regulation is established that limits polar compounds to 25% and polymer content to 12%.

6.6.5 Effect on Protein Content of Food

The nutritional value of food proteins is defined by a combination of quality and quantity. Quality represents the consumption of the characteristic content of the protein by the living organism; quantity is the amount of protein content present in the food. Reduction of protein content and destruction of some amino acids caused by heat treatment during DFF affect the quality of the protein content (Henry, 1998). On the other hand, after the DFF process, the protein content can be increased by frying the product in grass carp oil (Zhang et al., 2013).

In some studies, DFF products were observed to have higher protein content than those that were left untreated. Experimental studies found that samples fried in palm oil have the lowest amino acid content (Oluwaniyi et al., 2010). Decreases in lysine content of up to 17% and 25% were found when fish oil was used for continuous frying for 48 h in DFF operations, due to interactions between the carbonyl compounds and the amino group of lysine.

Very little effect on the amino acid content has been observed during frying. Changes occur in selected fish (arginine, aspartic acid, serine, glutamic acid, proline, glycine, alanine, lysine, histidine, threonine, valine, methionine, leucine, isoleucine, phenylalanine, cysteine, and tyrosine) (Ismail and Ikram, 2004). An experiment was carried out to determine changes in amino acid content on raw, cooked, and fried fish, but no significant differences were found. The frying mechanism can be said to have the same effect as the dehydration mechanism. Theoretically, there is no agreement over the loss of essential amino acids. However, loss of lysine is supposed to occur due to its involvement in the Maillard reaction as the first amino acid during the frying process.

6.6.6 Effect on the Mineral Content of Food

In DFF operations, mineral components show several changes, due to their solubility in water. They are soluble in trace amounts, due to which the effects of frying on minerals are almost negligible. The weight of fried food decreases due to loss of moisture content. Most of the minerals are nonvolatile, increasing the wet weight on the uptake of frying oil. Moderate decreases in mineral content can be observed in dried foods (Pokorny, 1999).

Ozeren and Ersoy (2009) found that at high temperature (165–185°C), and with a short cooking time, minerals are relatively preserved by frying. They also concluded that DFF of fish increases the concentration of minerals i.e., sodium, potassium, phosphorus, magnesium, iron and zinc, due to concentration effects.

Asghari et al. (2013) studied rainbow trout fillets, where minerals were observed to increase significantly during frying.

An experiment was carried out to determine the effect of frying frozen seafood on mineral content and heavy metals. The heavy metal macro and micro elements content was found to have increased during the DFF operation (Czech and Stachyra, 2013).

6.6.7 Changes in Vitamin Content and Degradation of Antioxidants

Most vitamins are volatile and are heat, temperature and oxidation sensitive. Enzymatic oxidation and long frying times are the main reasons for the loss of vitamins (Somsub et al., 2008). Owing to their sensitivity, a high amount of vitamin loss occurs through oxidation. Retention of vitamin content is very high in the temperature range 70–90°C, but because the temperature is very high in DFF operations, high loss of vitamins takes place (Pokorny, 1999). Oxidation of vitamins depends on internal temperature and frying procedure. Vitamins, especially vitamin C, are thermo-sensitive. In the B group, riboflavin, thiamine, niacin, and B_6 are the most frequently affected (Reda, 2004).

Some carotene oxidation products have been identified in heated palm oil, indicating that the loss of beta-carotene during frying is probably partly due to the oxidation of the compound retinol. Carotenoids and tocopherols are destroyed, changing the flavor and color of the oil. However, preferential oxidation of tocopherols has a protective (antioxidant) effect, which is particularly important, since the majority of frying oils are of vegetable origin, containing large amounts of unsaturated, rapidly oxidized fats (Fellows, 2006).

Barrera-Arellano et al. (2002) found that in oils with different degrees of unsaturation, degradation was significantly higher when tocopherols were absent.

Kourimskaand Gordon (1995) reported that after four to five frying operations, α-tocopherol content was more quickly lost than other tocopherols of canola oil, which were reduced by 50%.

6.7 UPTAKE OF FAT IN DFF FOODS

Foods that are deep-fat fried have visually attractive characteristics, excellent flavor, and crispy crust. They contain almost 45% fat (Funami et al., 1999). Water migrates from the inner core of the food to the crust during frying and is evaporated, leaving a permeable surface. The amount of fat uptake in the voids left by the vapor is determined by the moisture content of the food (Pinthus et al., 1995).

Health hazards related to high fat consumption include raised cholesterol level, high blood pressure, and heart disease. Consumer interest in low-fat food is increasing significantly. Fat makes up almost one-third of the content of fried food products by weight. Recent improvements are based on the modification of frying procedures, the frying medium, and the properties of the surface. Appropriate techniques include conventional shaking, draining, cooking at suitable frying time and temperature (Mellema, 2003). Modification of viscosity, monitoring of degraded frying oil, and the application of batters or coatings (use of hydrocolloids) are some of the alternative techniques for reducing fat uptake while preventing loss of nutritional properties during processing.

6.8 NUTRITIONAL ASPECTS OF FRIED FOOD

The composition of several foods changes when fat penetrates the food and modifies its structural as well as textural properties, just like the process of chromatography. In thermal, industrial or culinary processes there is always concern about the effect of various factors, including deep frying, on nutritional value. The numerous changes in nutritional properties of DFF food products depend upon factors such as composition of the frying fat, size, shape, and texture, and conditions such as temperature and frying time (Varela, 1998). Many undesirable effects may occur simultaneously such as loss of nutrients and vitamins during the frying operation (Fillion and Henry, 1998).

6.9 EFFECT OF DEEP-FAT FRYING ON NUTRITIONAL PROPERTIES OF FOOD

Consumption of any food product at an extreme level is injurious to health. In fried food products, the amount of oil/fat absorbed during the process is the main quality parameter. In order to reduce the total dietary fats, trans and saturated fatty acids, some nutritional recommendation has been made and point out to validate and limit the intake of oil in large amount. It can prevent degradation of oil from which toxic molecules generates either in frying oil or fried in fried products during DFF process (Bouchon and Pyle, 2004).

6.9.1 Effect on Human Health of Consumption of Fried Food

Fried foods are very popular with consumers, mainly because of the pleasant taste and crunchiness that develops during frying. The link between high consumption of fat and heart disease, high blood pressure, diabetes, and cancer is well documented (Saguy and Dana, 2003). The effect of the concentration of high-density and low-density lipoproteins has also been extensively researched (Ascherio and Willett, 1997).

Trans fatty acids are produced during hydrogenation for thermal stability of frying oil and in high thermal processing, such as DFF (Choe and Min, 2007). Consumer trends moving towards healthier, lower-fat food has led to development of products with a reduced oil content. But in developed countries, enormous quantities of high-fat foods are still consumed, with fat content ranging from 5% in frozen French fries up to 40% in potato chips (Dobraszczyk et al., 2006). The contribution of total saturated and trans fatty acids intake mainly comes from fried foods, so high-quality oil must be recommended to maintain both product quality and consumer health (Minihane and Harland, 2007).

Numerous research studies point to cell damage and carcinogenic properties associated with food products treated in highly oxidized conditions and at high temperature. HNE (4-hydroxy-2-trans-nonenal), a product of lipid peroxidation, has shown some mutagenic and toxic properties. Peroxidation of secondary lipids that are highly reactive with proteins, and the nucleic acids DNA and RNA, links them to severe liver disease, Alzheimer's and atherosclerosis (Seppanen and Csallany, 2006).

Toxic molecules can be generated inside food products during frying. In April 2002, Swedish scientists discovered a high level of acrylamide in snacks like potato crisps and French fries, leading the WHO to warn of possible links to carcinogens and hazards to human health. Mottram et al. (2002) state that acrylamide substances generated due to frying food at temperatures above 120°C causes a reaction between reducing sugar and amino acids. While the WHO has not yet called for reduction of high levels of acrylamide in food products, implementation of proper strategy for controlling acrylamide content in fried food is one of the emerging aspects. (Dueik and Bouchon, 2011; Besaratinia and Pfeifer, 2004; Koyama et al., 2006).

6.9.2 Product Quality and Shelf Life

The attractive appearance, texture, aroma, flavor, and sensory-related properties of fried products depend on both raw material and frying conditions. The frying medium itself is one of the important ingredients, especially for snacks and chips.

The nutritional content of fried products depends on the quality of oil that is absorbed by the food. When oil deterioration occurs, compounds are produced that are uniformly distributed in the fryer (Guillaumin, 1988). Oil that is extremely deteriorated not only renders the quality of the fried food product unacceptable; it also damages the product in storage conditions, causing bad odors and adversely affecting health. Oil undergoes oxidation and hydrolysis in storage but the reaction rate will be slower than during frying. Lipase enzymes are the main cause of hydrolytic alterations in the presence of moisture. Maintaining low moisture and inactivating enzymes during storage can prevent hydrolytic rancidity.

Oxidation of fried food is caused by the presence of oxygen, light, and increasing temperature. Oxidative rancidity can be prevented by cold storage without oxygen and light, or with the addition of artificial antioxidants. Foods containing a high amount of unsaturated fatty acid are unstable and deteriorate faster than those containing saturated fatty acids. The addition of moisture accelerates rancidity during storage, leading to loss of crispiness. Packaging material that contains a moisture barrier helps to prevent the absorption of moisture during storage and consequent deteriorative changes. Shelf life is extended if light is not admitted. Insertion of nitrogen gas inside the packaging stabilizes the treated product for longer.

6.10 HYDROCOLLOIDS FOR COATING DURING DFF

Hydrocolloids, which are regarded as multifunctional food additives, are useful for coating during DFF. They are mainly added during frying to control absorption of oil and to improve the textural and physical properties of food (binding capacity of water, viscosity of oil, and emulsion stability). Some hydrocolloids are reported as long chain polysaccharides that help reduce oil absorption during DFF. Fiszman and Salvador (2003) state that the application of hydrocolloids may take two forms: as an edible coating of aqueous solution on the food to be fried; and as an ingredient of the batter used in battered or breaded foods.

Edible material applied to the surface of the food as a protective coating acts as a barrier to infiltration of oxygen, moisture and solutes to the food product (Mchugh and Senesi, 2000). The coating may be on solid foods or between foods with thin edible layers (Krochta and Mulder-Johnston, 1997). There is a difference between edible coating and edible film. Edible film is prepared separately by casting a biopolymer solution forming a film which is later applied to the food, while edible coating is directly applied to the food material in the form of an emulsion or suspension solution which later becomes a film. Neither edible film nor edible coating are regarded as traditional packaging materials in the food industry but they nevertheless contribute to the preservation purposes of packaging.

Edible coatings are required to pass food safety standards because they are consumed along with the food. One of the main reasons for using edible coatings and films is to prevent water loss. They also modify product appearance, create texture, protect ingredients, increase shelf life, improve taste, prevent migration of core constituents, and maintain structural integrity (Fiszman and Salvador, 2003). They may also provide additional nutrients, enhance sensory attributes and include quality-enhancing antimicrobials. In addition, the use of edible coating and films is economically advantageous and reduces packaging (Campos et al., 2010). A very common use of coatings in DFF foods is breading and batters, the former comprising breadcrumb or cracker granules applied dry to form a suitable texture, and the latter a mixture of water, flour, seasoning, hydrocolloids, starch, flavoring, and other ingredients. Before the food is fried it is dipped into the solution of a thick liquid or semi-liquid system to form a layer. Two different batters are used in DFF, adhesive/interface and puff/tempura batter (Loewe, 1990). Adhesive batters act as a gluing agent that helps the breading material to be coated uniformly on the surface of the food, whereas puff/tempura batter itself serves as the exterior coating of the food material and is usually made from chemically leavened liquid dough.

A number of factors affect the functional properties of the batter system, including water-holding capacity, thermal properties, and viscosity. It is an extremely flexible system that helps in changing the malleability of the product during its development. Several recent studies have established a relationship between batter formulations, coating system properties, and quality characteristics such as texture, appearance, and adhesion power (Xue and Ngadi, 2006). The two main purposes of using hydrocolloids in breading/batter systems are to maintain viscosity and enhance water-holding capacity (Rimac-Brncic et al., 2004). This results in the development of network structures that minimize mass transfer during frying, which helps reduce the absorption of oil by the food product during DFF (Dogan et al., 2005).

6.11 GUMS FOR COATING DURING DEEP-FAT FRYING

Different types of gums exist in nature, including exudate gums, seed gums, and microbial fermentation gums. Xanthan gum is a type of microbial fermentation gum that provides a uniform coating and can be useful in frying operations. Chitosan, a natural polysaccharide developed from the alkaline deacetylation of chitin β-(1→4)-linked D-glucosamine, is an example of animal source coating. Gennadios et al. (1994) found that the exoskeletons of crustaceans, such as shrimps, crayfish, and crabs, produce chitosan. Animals are a rich source of proteins such as casein and whey protein; soy protein, zein and gluten can be obtained from plants. Proteins are able to form coatings or thin-film layers that act as a barrier and their ability mainly depends on their molecular weight, electrical properties, conformation, thermal stability and flexibility (Vergas et al., 2008). There has been less research on proteins than on polysaccharides. The performance of hydrocolloids is determined mainly by their ability to interact with other food components. The digestibility of hydrocolloids and their inclusion in batter coatings significantly affects textural and rheological properties of products being deep-fat fried (Meyers and Grazela, 1990).

6.12 CONCLUSION

Consumer demand for fried food is high and in general consumers are not willing to sacrifice its taste advantages for healthier food. This chapter has discussed the practical and theoretical aspects of deep-fat frying such as oil uptake, heat and mass transfer mechanisms, and analytical aspects of fat degradation. The literature concludes that consumption of large amounts of fat is damaging to health and contributes to severe diseases related to the heart, high cholesterol, and high blood pressure. However, some research suggests that coating of fried products will help reduce oil uptake, so there is lower absorption of oil content into food being deep-fat fried. Packaging materials and shelf stability products should be checked before final packaging and storage. For consumer safety and health, it is important that adequate processing operation, quality control, and regular monitoring of fried products is carried out so that quality, quantity, and nutritional attributes can be retained.

REFERENCES

Aguilera, J.M. and H. Gloria. 1997. Determination of oil in fried potato products by differential scanning calorimetry. *J. Agric. Food Chem.* 45: 781–785.

Asghari, L., F. Zeynali, and M.A. Sahari. 2013. Effects of boiling, deep-frying, and microwave treatment on the proximate composition of rainbow trout fillets: Changes in fatty acids, total protein and minerals. *J. App. Ichthy.* 29: 847–853.

Barrera-Arellano, D., M.V. Ruiz-Méndez, J. Velasco, G. Márquez-Ruiz, and M.C. Dobarganes. 2002. Loss of tocopherols and formation of degradation compounds at frying temperatures in oils differing in unsaturation degree and natural antioxidant content. *J. Sci. Food Agri.* 82: 1696–1702.

Baumann, B. and E. Escher. 1995. Mass and heat transfer during deep fat frying of potato slices: I. Rate of drying and oil uptake. *Lebensmittel-Wissenschaft Tech.* 28: 395–403.

Besaratinia, A. and G. Pfeifer. 2004. Genotoxicity of acrylamide and glycidamide. *J. Natl. Cancer Inst.* 96: 1023–1029.

Blumenthal, M.M. 1991. A new look at the chemistry and physics of deep-fat frying. *Food Tech.* 45: 68–71.

Bouchon, P. and Aguilera, J.M. 2001. Microstructural analysis of frying of potatoes. *Inter. J. Food Sci. Tech.* 36: 669–676.

Bouchon, P., J.M. Aguilera, and D.L. Pyle. 2003. Structure oil absorption relationships during deep-fat frying. *J. Food Sci.* 68(9): 2711–2716.

Bouchon, P., P. Hollins, M. Pearson, D.L. Pyle, and M.J. Tobin. 2001. Oil distribution in fried potatoes monitored by infrared microspectroscopy. *J. Food Sci.* 66: 918–923.

Bouchon, P. and D.L. Pyle. 2004. Studying oil absorption in restructured potato chips. *J. Food Sci.* 69: E115–E122.

Bouchon, P. and D.L. Pyle. 2005. Modelling oil absorption during post-frying cooling: Solution of the mathematical model, model testing and simulations. *Trans. Inst. Chem. Eng. C Food Biopro.* 83: 253–260.

Campos, C.A., L.N. Gerschenson, and S.K. Flores. 2010. Development of edible films and coatings with antimicrobial activity. *Rev. Food Biopro. Tech.* 4: 849–875.

Choe, E. and D. Min. 2007. Chemistry of deep-fat frying oils. *J. Food Sci.* 72: R77–R86.

Chung, J., J. Lee, and E. Choe. 2004. Oxidative stability of soybean during frying of flour dough. *J. Food Sci.* 69: 574–576.

Cuesta, C., F.J. Sanchez-Muniz, C. Garrido-Polonio, S. Lopez-Varela, and R. Arroyo. 1993. Thermo-oxidative and hydrolytic changes in sunflower oil used in frying with a fact turnover of fresh oil. *J. American Oil Chem. Soc.* 70: 1069–1073.

Czech, A. and K. Stachyra. 2013. Effect of processing treatments (frozen, frying) on contents of minerals in tissues of "frutti di mare". *Inter. J. Food Sci. Tech.* 48: 238–245.

Deman, J.M. 1999. *Principles of Food Chemistry*, 3rd edn. Gaithersburg, MD: Aspen Publishers, Inc.

Dobraszczyk, B.J., P. Ainsworth, S. Ibanoglu, and P. Bouchon. 2006. Baking, extrusion and frying. In: J.G. Brennan (ed.), *Food Processing Handbook*. Weinheim: Wiley-VCH, pp. 237–290.

Dogan, F.S., S. Sahin, and G. Sumnu. 2005. Effects of soy and rice flour addition on batter rheology and quality of deep-fat fried chicken nuggets. *J. Food Eng.* 71(1): 127–132.

Dueik, V. and P. Bouchon. 2011. Vacuum frying as a route to produce novel snacks with desired quality attributes according to new health trends. *J. Food Sci. 76*(2): E188–E195.

Farkas, B.E. 1994. Modeling immersion frying as a moving boundary problem. PhD dissertation, University of California, Davis.

Fellows, P.J. 2006. *Food Processing Technology, Principles and Practice*. Sao Paulo, Brazil: Artmed.

Fillion, L. and C.J. K. Henry. 1998. Nutrient losses and gains during frying: A review. *Inter. J. Food Sci. Nutri.* 49: 157–168.

Fiszman, S.M. and A. Salvador. 2003. Recent developments in coating batters. *Tren. Food Sci. Tech.* 14(10): 399–407.

Funami, T., M. Funami, T. Tawada, and Y. Nakao. 1999. Decreasing oil uptake of doughnuts during deep-fat frying using curdlan. *J. Food Sci.* 64(5): 883–888.

Gennadios, A., T.H. Mchugh, G.L. Weller, and J.M. Krochta. 1994. Edible coatings and films-basedonproteins. In: J.M. Krochta, E.A. Baldwin, and M.O. Nisperos-Carriedo (eds.), *Edible Coatings and Films to Improve Food Quality*. Lancaster, PA: Technomic Publishing Co., Inc., pp. 201–277.

Gertz, C. 2005. *Quality Control and Safety of Cooking Oils*. Frankfurt: Workshop Deep-Frying.

Giese, J. 1996. Fats, oils, and fat replacers. *Food Tech.* 50(4): 78–84.

Guillaumin, R. 1988. Kinetics of fat penetration in food. In: G. Varela, A.E. Bender, and I.T. Morton (eds.), *Frying of Food: Principles, Changes, New Approaches*. Chichester, UK: Ellis Horwood, pp. 82–89.

Henry, C.J.K. 1998. Impact of fried foods on macro-nutrient intake, with special reference to fat and protein. *Fat Oil.* 49: 336–339.

Hubbard, L.J. and B.E. Farkas. 2000. Influence of oil temperature on heat transfer during immersion frying. *J. Food Proc. Preser.* 24: 143–162.

Ismail, A., and E.H. K. Ikram. 2004. Effects of cooking practices (boiling and frying) on the protein and amino acids contents of four selected fishes. *Nutri. Food Sci.* 34: 54–59.

Kourimska, L. and M.H. Gordon. 1995. The effects of antioxidants on changes in oils during heating and deep frying. *J. Sci. Food Agri.* 68: 347–353.

Krochta, J.M. and C.D. Mulder-Johnston. 1997. Edible and biodegradable polymer films: Challenges and opportunities. *Food Tech.* 51: 61–74.

Loewe, R. 1990. Ingredient selection for batter systems. In: K. Kulp and R. Loewe (eds.), *Batters and Breadings in Food Processing*. St. Paul, MN: American Association of Cereal Chemists, Inc., pp. 11–28.

Mariscal, M. and P. Bouchon. 2008. Comparison between atmospheric and vacuum frying of apple slices. *Food Chem.* 107(4): 1561–1569.

Mchugh, T.H. and E. Senesi. 2000. Apple wraps: A novel method to improve the quality and extend the shelf life of fresh-cut apples. *J. Food Sci.* 65: 480–485.

Mellema, M. 2003. Mechanism and reduction of fat uptake in deep-fat fried foods. *Inter. J. Food Sci. Tech.* 14: 364–373.

Meyers, M.A. and A.A. Grazela. 1990. Functionality of hydrocolloids in batter coating systems. In: K. Kulp and R. Loewe (eds.), *Batters and Breading in Food Processing*. St. Paul, MN: American Association of Cereal Chemists, Inc., pp. 117–138.

Minihane, A.M. and J.I. Harland. 2007. Impact of oil used by the frying industry on population fat intake. *Criti. Rev. Food Sci. Nutr.* 47: 287–297.

Mottram, D.S., B.L. Wedzicha, and A.T. Dodson. 2002. Food chemistry: Acrylamide is formed in the Maillard reaction. *Nature*, 419: 448–449.

Nawar, W.W. 1985. Chemistry of thermal oxidation. In: D.B. Min and T.H. Smouse (eds.), *Flavour Chemistry of Fats and Oils*. Champaign, IL: American Oil Chemists Society, pp. 39–60.

Ngadi, M. and J. Xue. 2009. Food frying: Modifying the functional properties of batters. In: J. Ahmed, H.S. Ramaswamy, S. Kasapis, and J.I. B. Boye (eds.), *Novel Food Processing: Effects on Rheological and Functional Properties*. Boca Raton, FL: CRC Press, pp. 437–457.

Oluwaniyi, O.O., O.O. Dosumu, and G.V. Awolola. 2010. Effect of local processing methods (boiling, frying and roasting) on the amino acid composition of four marine fishes commonly consumed in Nigeria. *Food Chem.* 123: 1000–1006.

Ozeren, A. and B. Ersoy. 2009. The effect of cooking method on mineral and vitamin contents of African catfish. *Food Chem.* 115: 419–422.

Peers, K.E. and P.A.T. Swoboda. 1982. Determination of sunflower seed oil under simulated frying conditions and during small scale frying of potato chips. *J. Sci. Food Agri.* 33: 389–395.

Pinthus, E.J., Weinberg, P., and Saguy, I.S. 1995. Oil uptake in deep fat frying as affected by porosity. *J. Food Sci.*, 60(4): 767–769.

Pokorny, J. 1999. Changes in nutrients at frying temperatures. In: D. Boskou and I. Elmadfa (eds.), *Frying of Food*. Boca Raton, FL: CRC Press.

Reda, S.Y. 2004. Comparative study of vegetable oils subjected to the thermal stress. Master's dissertation. Universidade Estadual De Ponta, Grossa, Brazil.

Rimac-Brncic, S., V. Lelas, D. Rade, and B. Simundic. 2004. Decreasing of oil absorption in potato strips during deep fat frying. *J. Food Eng.* 64(2): 237–241.

Saguy, S. and D. Dana. 2003. Integrated approach to deep fat frying: Engineering, nutrition, health and consumer aspects. *J. Food Eng.* 56: 143–152.

Sanchez-Muniz, F.J., C. Cuesta, M.C. Lopez-Varela, M.C. Garrido-Polonio, and R. Arroyo. 1993. Evaluation of the thermal oxidation rate of sunflower oil using various frying-methods. In: T.H. Applewhite (ed.), *Proceedings of World Conference on Oilseedand Technology and Utilization* (pp. 448–452).

Seppanen, C.M. and A. Csallany. 2006. The effect of intermittent and continuous heating of soybean oil at frying temperature on the formation of 4-hydroxy-2-trans-nonenal and other a-, b-unsaturated hydroxyaldehydes. *J. Am. Oils Chem. Soc.* 83: 121–127.

Singh, R.P. 1995. Heat and mass transfer in foods during deep-fat frying. *Food Tech.* 49: 134–137.

Somsub, W., R. Kongkachuichai, P. Sungpuag, and R. Charoensiri. 2008. Effects of three conventional cooking methods on vitamin C, tannin and myo-inositiol phosphates contents in selected thai vegetables. *J. Food Compre. Anal.* 21: 187–197.

Stevenson, S.G., M. Vaisey-Genser, and N.A. M. Eskin. 1984. Quality control in the use of deep-frying oils. *J. American Oil Chem. Soc.* 61: 1102–1108.

USDA. 2013. United States Department of Agriculture. Retrieved from: http://ndb.nal.usda.gov/ndb/foods/show/6572.

Varela, G. 1988. Current facts about the frying of food. In: G. Varela, A.E. Bender, and I.D. Morton (eds.), *Frying of Food: Principles, Changes, New Approaches*. Chichester: Ellis Horwood, pp. 9–25.

Vergas, M., C. Pastor, A. Chiralt, D.J. Mcclements, and C. Gonzalez-Martinez. 2008. Recent advances in edible coatings for fresh and minimally processed fruits. *Crit. Rev. Food Sci.* 48: 496–511.

Vitrac, O. 2000. Caractérisationexpérimentale et modélisation de l'opération de friture. PhD thesis, Ecole NationaleSupérieure des Industries Agricoles et Alimentaires, Massy, France.

Vitrac, O., G. Trystram, and A.L. Raoult-Wack. 2000. Deep-fat frying of food: Heat and mass transfer, transformations and reactions inside the frying material. *European J. Lip. Sci. Tech.* 102: 529–538.

Xue, J. and M. Ngadi. 2006. Rheological properties of batter systems formulated using different flour combinations. *J. Food Eng.* 77(2): 334–341.

Zhang, J., D. Wu, D. Liu, Z. Fang, J. Chen, Y. Hu, and X. Ye. 2013. Effect of cooking styles on the lipid oxidation and fatty acid composition of grass carp (Ctenopharynyodonidellus) fillet. *J. Food Biochem.* 37: 212–219.

Ziaiifar, A.M., N. Achir, F. Courtois, I. Trezzani, and G. Trystram. 2008. Review of mechanisms conditions and factors involved in the oil uptake phenomenon during the deep fat frying process. *Inter. J. Food Sci. Tech.* 43: 1410–1423.

7 Crystallization
Fundamentals, Current Research, and Role in Product Development

Winny Routray
Department of Food Process Engineering, National Institute of Technology Rourkela, Rourkela, Odisha, India

CONTENTS

7.1 INTRODUCTION

Crystallization is one of the major processing technologies that regulate quality, shelf life and consumer acceptability of many food commodities. The storage stability and appeal of high-value products such as ice cream and chocolate are determined by crystallization techniques and crystal properties. The process affects both physical and sensory properties, including appearance, hardness, feel, taste, texture, and dispersion stability (Miyagawa et al., 2016; Payne et al., 2019). In certain popular products, crystallization determines the end of storage life, such as lactose crystallization in ice cream, and sugar crystallization and fat blooms (Sonwai and Rousseau, 2010) in chocolate. In some commodities changes in quality can be correlated with the crystallization of a constituent. In the case of mayonnaise, storage at lower temperatures is a common practice that can lead to destabilization caused by lipid crystallization (Miyagawa et al., 2016). Certain high-melting compounds in the oil or fat are also crystallized to eliminate them from the product, so that physical properties of the final product, such as texture, can be targeted (Abd. Rashid et al., 2012).

This field of food engineering is constantly evolving and advanced techniques are being developed to control the process of crystallization in terms of time, mechanism, crystal content, and size, shape, and arrangement of crystals. Traditional and novel food items with increased biological functionality and sensory acceptance are also increasingly being prepared. Furthermore, modern techniques for quantification and modeling of the process are also being increasingly explored. This chapter focuses on the research conducted in these areas over the past two decades, along with research that might be of increasing interest or emphasis in future. Discussion of the different crystallization mechanisms is beyond the scope of this chapter, but these have been widely reviewed elsewhere (Ferguson et al., 2012; Mathew Thomas and Lakerveld, 2019; Wang et al., 2017) (Table 7.1).

7.2 PRINCIPLES OF CRYSTALLIZATION

7.2.1 Principles

The crystallization process can be divided into four phases: (a) transitional phase in liquid, where a supersaturated stage is achieved along with a driving force for crystallization; (b) nucleation, which starts when the energy barrier for crystallization is overcome and molecules in liquid state join together to form crystals; (c) crystal growth phase, which commences as soon as the nuclei are stabilized, when crystals grow through aggregation of other molecules present in the liquid onto the crystal lattice, followed by reduction in the driving force of super saturation and are then stopped by some kinetic constraint or continue till the equilibration of the system (driving force reaches zero and maximum phase volume of the crystal is achieved); and (d) recrystallization or ripening, which generally occurs during storage to attain overall equilibrium (Hartel, 2013), in which smaller crystals can be dissolved followed by growth of larger ones (Payne et al., 2019). In certain cases, nucleation can be further divided into primary and secondary nucleation. Secondary nucleation has been described as a phase where the previously formed nuclei catalyze the nucleation and formation of more of nuclei. This may occur at a higher rate than primary nucleation and is affected by various parameters, including super saturation, cooling, agitation, and the presence of impurities (Kiani and Sun, 2011).

Nucleation is either initiated spontaneously (by overcoming the thermodynamic barrier) or by introducing a seed, following which crystal growth occurs. Depending on the case and the biomaterials, heating followed by evaporation and concentration or lowering of temperature (Crafts, 2007) lead to crystallization. The thermodynamic driving force has been described as the "chemical potential difference between the solution and the crystal". However, the temperature difference between the actual level and melting point and/or the concentration difference between the real and the solubility concentration are used to specify the driving force (Payne et al., 2019).

TABLE 7.1
Books for Further Detailed Study and Reference

Title of Book	Brief Summary of Content	Authors/Editors of Books/Book Chapter
Crystallization	Includes all aspects of crystallization including mechanism, industrial-scale instruments, and crystallizer design and operation.	Mullin (2001)
Crystallization technology handbook	Detailed explanation of crystallization phenomenon and corresponding chemical engineering processes, lab-scale principles, and large-scale crystallizers.	Mersmann (2001)
Crystallization and solidification properties of lipids	Provides insight on advanced techniques for the study of fat crystallization, including major product examples such as chocolate, oils, and other emulsions	Widlak et al. (2001)
Crystallization processes in fats and lipid systems	Comprehensive overview of fat crystallization mechanism and kinetics, phase transitions, and different fat networks in generic products and derivatives.	Garti and Sato (2001)
Additives and crystallization processes from fundamentals to applications	Comprehensive account of the effect of application of different additives on crystallization.	Sangwal (2007)
Chemical deterioration and physical instability of food and beverages	Includes a chapter on crystallization in foods and food quality deterioration.	Bund and Hartel (2010)
Crystallization: basic concepts and industrial applications	Presents basic crystallization mechanism; comprehensively discusses characterization techniques, basics of industrial crystallization, techniques employed at different stages/unit operations, and effect of additives and impurities.	Beckmann (2013)
Non-equilibrium states and glass transitions in foods processing effects and product-specific implications	Detailed discussion of non-equilibrium states and glass transitions encountered during crystallization occurring in different processing techniques and various product preparations.	Bhandari and Roos (2016)
Crystallization of lipids fundamentals and applications in food, cosmetics, and pharmaceuticals	Wide variety of content includes lipid crystals in different lipid networks that are important components of food, cosmetics, and pharmaceuticals.	Sato (2018)
Confectionery science and technology	Summarizes different aspects of sugar crystallization and mechanisms involved in preparation of confectionery and sweets.	Hartel et al. (2018)
Reference module in food science	Includes a chapter recent advances in different aspects of crystallization in foods, emphasizing mechanism, quantification of crystallization, and new methods involved in control of crystallization in food commodities.	Payne et al. (2019)
Handbook of industrial crystallization	Includes chapters on application of molecular and population modeling in crystallization, different processes of crystallization, and industrial crystallizers.	Myerson et al. (2019)

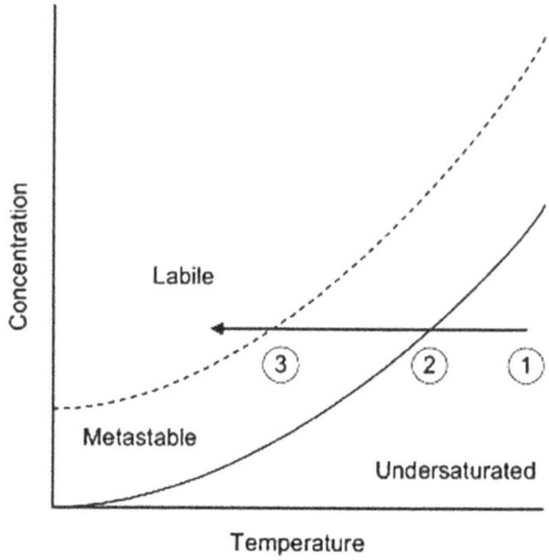

FIGURE 7.1 Super saturation, under saturated, and metastable zone width in cooling crystallization. *Source*: Crafts (2007)

Apart from the thermodynamic barrier, resistance to mass transfer (interactions between the molecules and molecular mobility) in the solution can be major constraints on the formation of nuclei, preventing crystallization (even after application of significant thermodynamic force). In solubility and saturation curves (Figure 7.1), a metastable region is defined, where initiation of crystallization is difficult, as the aggregates formed subsequently disperse, preventing nucleation. This region exists between super saturation and saturation curves and varies with system properties, complexity, and flexibility of the solute molecules (Crafts, 2007). The following sections discuss some of the phenomena that play an important role in crystallization and have been topics of recent research. Detailed generic discussion of the mechanism and techniques of crystallization for further reference and study can be found in previous publications, which are summarized in Table 7.1.

7.2.2 Behavior of Components in Solution

The compounds or biomolecules present in a solution interact with each other, leading to corresponding solution behavior and effect on nucleation and crystal growth. In recent decades, a two-step model has led to a better explanation of nucleation than that provided by classical nucleation theory (Erdemir et al., 2009), as per which the crystalline nucleus shows up inside previous metastable bunches of size several hundred nanometers, which comprise of thick fluid and are suspended in the solution. The two-step nucleation model elucidates that in cases with large density fluctuations around the critical point, the pathway for nucleation can change significantly. In these cases, highly disordered liquid droplets, and subsequently a crystalline nucleus of a size beyond critical dimensions, can form. Consequently, the reduction in the free energy barrier for crystallization is attributed to proximity to the critical point, leading to increased nucleation rates. The intermediate phase may present as high-concentration

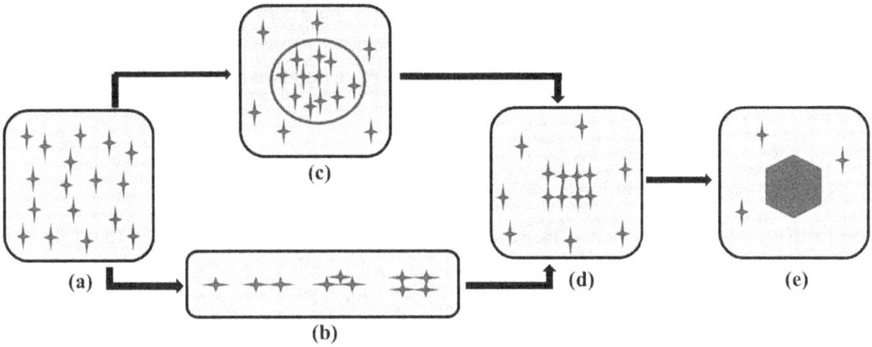

FIGURE 7.2 Pathways for the formation of crystals from solution: (a) supersaturated solution; (b) cluster of solute molecules, ordered and subcritical; (c) dense antecedent with liquid-like solute cluster; (d) ordered crystalline nuclei; (e) solid crystal. The classical nucleation model consists of steps (a), (b), (d) and (e). The two-step nucleation model pathway consists of steps (a), (c), (d) and (e). Figure based on Erdemir et al. (2009)

liquid, which has been described as a generic characteristic of biomolecules interacting through short-range interactions, as observed in proteins (Erdemir et al., 2009). Different pathways adopted during nucleation are shown in Figure 7.2.

Recently, interaction of carbohydrates with surrounding water molecules has been studied in detail, which has augmented the understanding of crystallization steps occurring in the corresponding carbohydrate solutions. In a study by Bensouissi et al. (2010), the metastable zone widths of aqueous solutions of sucrose, maltitol, xylitol, and mannitol were assessed, and larger metastable zone widths were found to correlate with higher possibilities of hydrogen bonds between the water and the corresponding solute present in individual solutions. Hence, the nucleation rate of the solute with narrower metastable zone width (maltitol) was higher than that of one with broader metastable zone width (xylitol), which was attributed to the lower affinity to water, higher diffusivity, and the presence of identical hydroxyl group orientation in both crystalline and solution states. Similar observations have also been reported in the case of other disaccharides such as trehalose and sucrose, compared with maltitol (Gharsallaoui et al., 2008). In this case, the narrowest metastable zone width was reported for maltitol and the widest for trehalose, which affected the nucleation rate and was attributed to the extent of affinity of the compounds to water, subsequently affecting the nucleation rate. However, it was also reported that hydration water should be removed and subsequently evacuated from the crystal integration surface to the bulk solution, for the crystallization of anhydrous disaccharides dissolved in aqueous media. These steps were deemed the necessary controlling steps for the crystallization of these anhydrous disaccharides.

Furthermore, based on the molecular dynamics simulations of the sucrose solution (Payne et al., 2019; Xia and Case, 2012a), the likely glycosidic bond conformation present in the solution has been reported to be the same as in the crystalline state. Even though the most abundant conformation has been identified as crystalline sucrose, it has also been found that the hydrogen bonding patterns break and reform in the solution, demonstrating the flexibility of the sucrose molecule. Similar observations have been reported for the range of concentrations between 0.1 and 4.0 M, when viscosity changes were observed with the change in concentration (Xia and Case, 2012b).

7.2.3 Phase Behavior and Crystallization Kinetics

Solubility curves or freezing point depression curves are used to demonstrate the equilibrium relationship between the liquid and crystalline phases. Binary phases can easily be represented by these curves, as binary phase diagrams include both equilibrium curves. However, as the complexity of food systems, including solid foods and beverages, are higher, individual components can contribute towards phase behavior. The solubility of one component could decrease when other components are introduced into the system, modifying the mechanism of crystallization as observed in a pure solution. Crystallization mechanisms observed in these systems are generally handled empirically (Hartel et al., 2018). Recently, the Flory–Huggins theory was used to predict the thermodynamic behavior of different types of sugars, including binary and ternary mixtures. The data obtained for individual sugars matched the data predicted by the model over a range of conditions. Quantitative mathematical predictions and corresponding methods can be employed for the better control of the process and can be used in the simulation of the mechanisms expected to be involved in newly developed products (Payne et al., 2019; Van der Sman, 2017).

Different models have been used to describe the crystallization process. The Avrami model, which describes the overall process, is often used for the quantification of crystallization kinetics under isothermal conditions (Abd. Rashid et al., 2012; Chai et al., 2018a, 2018b). The equation for the Avrami model, which is used to fit the observed data under isothermal conditions through nonlinear regression, can be expressed as Equation 7.1:

$$\frac{A_{(t)}}{A_{(\infty)}} = 1 - e^{-Kt^n} \tag{7.1}$$

where $A_{(t)}$ = the content of solid fat at crystallization time t,
$A_{(\infty)}$ = the content of limiting solid fat, when time approaches infinity,
K = the crystallization rate constant,
n = Avrami exponent that is sensitive to crystallization mechanism.

The parameters stated in Equation (7.1) consider both the processes of nucleation and the crystal growth rates. These parameters are also affected by the time for nucleation and dimensionality of the crystal growth (Chai et al., 2018a). In a study by Chai et al. (2018b), the increase in n value around a 10°C temperature correlated with a shift from one-dimensional growth to multi-dimensional growth in the network of fat crystals of palm kernel oil and coconut oil. The Avrami model has also been applied to the analysis of crystallization kinetics in a variety of samples, including individual samples of different types of fats and oils, and various blends (Abd. Rashid et al., 2012; Hubbes et al., 2018; Saberi et al., 2011; Wang et al., 2011; Xu et al., 2016a), sugars present in honey (Dettori et al., 2018), and recrystallized starch (Beck et al., 2011) under isothermal conditions. However, under non-isothermal conditions, other equations and models such as the Mo equation (in the case of monoglyceride organogels; Meng et al., 2014) have been employed. The Williams–Landel–Ferry kinetics equation has been used to describe crystallization kinetics in cases with the transition from amorphous to crystalline forms, along with the Avrami equation, where the crystallization rate represents the overall process and does not consider the differences between rates of nucleation and crystal growth (Das and Langrish, 2012). Furthermore, due to the complexity of fat crystallization, the kinetics of the process has also been explained using other models, including the Foubert model (relatively new) and the Gompertz model. In the case of fats with multiple crystallization peaks, the Gompertz model has been reported to fit the data better than the Avrami model (Chaleepa et al., 2010). The application of this model to fat crystallization has been attributed to the similarity between crystallization growth and microbial growth; however, as this model is empirical, it provides only limited understanding of the mechanical aspects of the crystallization process (Bayard et al., 2017).

An increasing recent trend has been the development of bioplastics for use in food packaging. One of the main challenges in the development of these products is increasing their strength. Recently, extensive research has been conducted on polylactic acid-based bioplastics, with the strength of these biomatrices having been increased in some cases through fortification with cellulose nanofibers and nanocrystals. Clarkson et al. (2019) employed conventional Avrami and Lauritzen–Hoffman analysis for the analysis of the crystallization behavior of polylactic acid-based bioplastic combined with cellulose nanocrystals and/or nanofibers with polyethylene glycol as plasticizer. They concluded that a synergistic effect existed between nanocellulose and polyethylene glycol, with the crystallization rate increasing in samples containing nanocellulose. The Avrami equation has also been employed in the study of crystallization behavior, kinetics and the effect of different parameters on other polylactic acid composites (Li et al., 2013).

Loss of the solute from the solution has also been further explored in terms of the kinetics of the process. In crystallization of sucrose from a sucrose solution through high-intensity ultrasound, the crystallization kinetics were interpreted with respect to the loss of the sucrose from the solution. In an attempt to increase crystal size or deposition of sucrose on the crystal seed, first-order kinetics were employed to explore the loss of solute from the solution during crystallization. The concentration of sucrose in real time was interpreted based on measured refractive indices (Singh et al., 2019).

7.3 ADVANCES IN THE STUDY OF FAT CRYSTALLIZATION

The study of fat crystallization in different commodities is one of the major challenges for scientists, manufacturers, and retailers involved in development of fat-based products, commercial large-scale production of commodities with crystallized components, and manufacture of crystallizers. The major challenge is the complexity of the crystalline structure of the fat, where the corresponding crystals do not behave as single crystal units, but rather as spherulite aggregates with large structures, in a wide range of sizes including macro, micro, meso, and molecular. It has recently been reported that during nucleation leading to the triacylglycerol crystal growth as spherulites, the primary particles were nanoplatelets derived from the corresponding lamellae, which subsequently aggregated to form a network of fat crystals through increase in super saturation along with decrease in the length and width of the nanoplatelets (Acevedo and Marangoni, 2015). The interaction between the various components of fats at the nanoscale, a major topic of research interest, should be further explored to understand the effects of different components on the properties of the newly formulated commodities.

In a recent study by Chai et al. (2018b), the effect of the non-triglyceride components of palm and coconut oil on the corresponding crystal structure were studied. It was observed that though the removal of these minor components had no effect on properties such as lipid composition and equilibrium solid fat content, certain other properties were affected. The slip melting point of oil with minor components was increased, compared to the fraction devoid of minor components. Based on the differential scanning calorimetry (DSC) profiles of the corresponding samples, it was deduced that the presence of minor components promoted the commencement of crystallization. In these oil samples, after the removal of minor components, nanoscale crystals increased in thickness with "no polymorphic transformation". In view of the polarized light micrographs, it was found that the change of coarser crystal structure to minuscule crystal structure happened in the microstructural networks of these oils due to the action of minor components; hence, when these minor components are removed, the crystallization mechanism is affected, leading to yet coarser texture (Chai et al., 2018b). Similar observations were also reported in a study on removal of indigenous minor components from hydrogenated palm kernel oil and coconut oil, where the crystal growth mechanism was affected by removal of these components (Chai et al., 2018a). Further study on the characterization and properties of these nanostructures could provide understanding of their role in the overall textural and chemical properties of other fats and fat-enriched products.

Traditional oils are also fortified with components such as diacylglycerols, which have been identified as novel functional lipids and can decrease postprandial lipid levels and fat accumulation in the body. These functionalities are attributed to the distinctive molecular structure of these compounds, and their beneficial metabolic characteristics. In a study by Xu et al. (2016b), blends of diacylglycerol-enriched palm oil and diacylglycerol-enriched palm olein were prepared, and the corresponding thermal properties, phase, and crystallization behavior were studied. It was observed that the crystallization rate increased and no spherulitic crystal growth was exhibited. Hence, an increasing volume of research has demonstrated the effect of concentration and composition on crystallization of different fat formulations under non-isothermal conditions (Daels et al., 2015).

7.4 FOOD PRODUCTS WITH CRYSTALS AS MAJOR COMPONENT

Traditional foods based on crystallization include butter and margarine. Chocolate, ice cream, and candy (fudge and fondant) are among the popular desserts and confectionery with sugar, water, and fat crystals. However, candy can be of two types, crystalline and non-crystalline. In the latter, crystallization of sucrose (predominant sugar) is undesirable and can be prevented either by addition of other sugars, including glucose and fructose, or slowed down using certain acids and fats. Furthermore, in a study by Chen et al. (2015), sucrose crystal growth rate decreased with the addition of corn syrup, along with a decrease in solubility temperature at higher levels of corn syrup, with no apparent effect on the glass transition temperature for the range observed. Crystallization is also an important processing method for the manufacture of original and candied dry fruits and dried fruit juice powders, which are becoming increasingly popular as they enable consumption of fruit throughout the year and provide instantly consumable nutritional commodities (de Oliveira Alves et al., 2010; Harnkarnsujarit and Charoenrein, 2011; Santhalakshmy et al., 2015; Zhao et al., 2015).

The microstructure of food components can be maintained by prolonging the non-crystalline state of the constituents. Crystallization of amorphous sugars can be delayed by fortification with other components such as starch, corn syrup solids, proteins, and carbohydrates. Fan and Roos (2016) investigated an amorphous sugar–whey protein combination system, where different properties including crystallization, calorimetric transition temperature, structural relaxations, relaxation times, and water sorption were observed. During this study, it was deduced that the crystallization mechanism and the crystalline forms of freeze-dried sugars with whey protein isolates were affected by the whey protein and trehalose. Further research can be conducted on the modification of crystallization behavior through compositional changes to achieve desirable properties, depending on the targeted properties of the commodity.

Chocolate is a major confectionery item in which sugar and fat crystallization controls quality during shelf life. During a study on crystallization of cocoa butter in cocoa powder, the process was observed under tumbling conditions, and it was concluded that crystallization of cocoa butter was affected by the attributes of the cocoa powder and the tempering process. Cocoa powder was reported to nucleate the cocoa butter. It was further observed that compared to bulk cocoa butter (only) samples, cocoa-butter samples with cocoa powder transitioned faster to polymorphic forms with higher stability. High-fat cocoas also demonstrated enhanced rates of crystallization compared to lower-fat cocoas, due to microstructural differences (Palmieri and Hartel, 2019). However, cocoa-butter emulsions containing water have recently been developed. In a study of crystallization in these emulsions, it was observed that the dispersed water droplets affected the phase transition of the fat phase (continuous) in two ways: increase in crystallization rate and enhancement of polymorphic evolution towards the formation of stable polymorphs. The nucleation mechanism did not change in the system; however, even at intermediate lower levels of super cooling, the emulsions crystallized faster. These observations were ascribed to the liquid-like layer of emulsifier (polyglycerol polyricinoleate) present around the water droplets, causing the cocoa-butter triglycerides to structurally rearrange, leading to the evolution of the polymorphs (Di Bari et al., 2017).

Ice crystallization is also an integral part of the ice-cream manufacturing process. Different mechanisms of ice crystallization during ice-cream production have been comprehensively discussed in a review by Cook and Hartel (2010). Ice crystallization and recrystallization play a major in the maintenance of frozen food product quality. However, due to the smaller difference between the melting points of large and small ice crystals, these crystals are extremely unstable during storage, which negatively affects the quality and shelf life of the product. Little is known about the thermodynamic aspects of this phenomenon, which deserves further investigation. However, recently, Ostwald ripening observed in the frozen foods has been analyzed using generic thermodynamic principles, when coarsening of ice crystals over time was predicted. It has been generally observed that these models could generate more accurate results with more data, which could then be used to predict shelf life (through modulation and prediction of crystal size over time) and to predict various ingredients (Payne et al., 2019; van Westen and Groot, 2018).

Despite the abundance of traditional recipes based on crystallization and the universal appeal of confectionery, businesses, food enthusiasts, and food scientists alike continue to working towards preparing innovative products. With increased interest in the preparation of novel and healthy confectionery items, traditional composition is being modified, which has led to crystallized products with properties different from traditional ones because of variation in the crystallization mechanisms. Food products are complex matrices with different components, all of which affect the crystallization process and its phases as well as product properties. Product properties can be further modulated by addition of components to obtain products with targeted desirable sensory and nutritional properties. In a study on the freezing of the Brazilian pequi fruit (*Caryocar brasiliense* Camb.), it was observed that pre-treatment using sucrose and alcohol improved the quality of the freeze-dried products and decreased the amount of collapsed structures. Pre-treatment of pulp with sucrose led to an amorphous and metastable product, whereas further application of alcohol pre-treatment augmented the drying process, subsequently improving the quality of the end product with no sign of collapse (de Oliveira Alves et al., 2010). Other reports on pre-treatments cover gas addition (Adhikari et al., 2018; Truong et al., 2017), thermal treatment (Alvarez et al., 2020), high-pressure processing (Zulkurnain et al., 2016), and compositional changes (Buscato et al., 2018; de Oliveira et al., 2015; Liu et al., 2019; Rizzo et al., 2015; Torbica et al., 2016), which can be further considered for improving and controlling crystallization mechanisms during the development of novel products.

7.5 RECENT ADVANCES IN CHARACTERIZATION AND PROCESSING TECHNIQUES

Apart from new product development, recent research has also concentrated on comprehensive and acute studies of the mechanism of crystallization and the development or application of non-traditional advanced methods for quantifying, controlling, and enhancing crystallization in food items. Important properties often achieved during crystallization include the targeted size, purity, morphology, and crystal structure (Erdemir et al., 2009).

7.5.1 Characterization Methods and Techniques

Traditional methods of characterization of food crystals include methods based on measurement of the refractive indices of solutions, microscopy, and X-ray techniques. All methods possess specific benefits and applicability. Recently developed advanced techniques are tuned/optimized for enhanced characterization and analysis of various crystals that demonstrate particular efficiency.

Fourier transform infrared (FTIR) spectroscopy has been widely employed to study structural features of compounds and the causes of differences in nucleation. In the case of biomolecules such as disaccharides, modifications in frequencies and the corresponding intensities around glycosidic bonds have been deduced in terms of the conformational flexibility of the molecules. Hydration stability of disaccharides has also been obtained based on the interpretation of FTIR spectra results. Combined attenuated total reflectance with Fourier transform infrared spectroscopy and Fourier transform near infrared spectroscopy have also been used to monitor in-situ concentration in batch-cooling crystallization processes (Kadam et al., 2010).

Microscopy techniques such as polarized light microscopy have been used for the determination of the morphology of crystallized samples. Cryogenic transmission electron microscopy has also been increasingly employed for the characterization of triacylglycerol crystal networks at nanoscale (Acevedo and Marangoni, 2010). Hot-stage microscopy with image analysis of crystallization, traditionally employed for observation of crystallization in pure fat, has also been used for the observation of crystallization in multicomponent natural fat (Harrison et al., 2016). Based on the thermal profiles obtained through methods such as DSC, transition of temperatures and corresponding heats are studied in terms of the procedures employed for melting and crystallization. X-ray diffraction is used to determine crystal polymorphism and crystalline domain size (Chai et al., 2018a). DSC and polarized light microscopy have been employed in combination for the study of the state behavior and crystal growth of thin films produced from a sucrose and corn syrup mixture (Chen et al., 2015). These methods are also used for the characterization of the crystal growth mechanism in other biomolecules such as fats (Chai et al., 2018b). Proton nuclear magnetic resonance (NMR) is used to determine fat content, based on the recognition and isolation of the proton signal in the fat molecules from all other proton-containing compositional sources present in a random food sample. During crystallization, this technique is used to monitor solid fat content changes with respect to time, which is often recorded at different temperature levels (Chai et al., 2018a; Payne et al., 2019).

X-ray micro-computed tomography is an efficient technique that has recently been applied to the non-destructive characterization of food microstructure. Three-dimensional images of the composition and density of a food sample can be obtained based on the absorbance and scattering of the passed X-rays. These images are created at very high speed, which makes them comparable to real-time measurements, a significant advantage over other methods. This technique has been employed for the characterization and quantification of the microstructure of various food products, including dairy, meat, fruits and vegetables, cereals, baked goods, and confectionery. It has also been used to determine other structural, compositional, and physical characteristics, salt and water concentrations, for characterization of various components of complex biomaterials, tracing the mechanism of microstructural evolution, and the evaluation of internal decay caused by post-harvest and processing operations (Schoeman et al., 2016). The technique has also been employed for the characterization of ice crystals in food biomatrices during the freezing process (Mousavi et al., 2007; Ullah et al., 2014). Other uses of the technique include the study of ice recrystallization mechanisms and their effect on the storage quality of salmon (Syamaladevi et al., 2012), the mechanism of the coarsening of air cells and ice crystals due to the thermal cycle in ice cream (Guo et al., 2017), and the characterization of the size of sucrose crystals and cracks in the structure of chocolate that could be caused by migration processes, leading to problems such as chocolate blooming (Reinke et al., 2016).

X-ray diffraction (XRD) has been used extensively for the structural characterization of various food components including sugars and fats (Jouppila et al., 1997), and for the identification of different crystalline phases present in a biological sample. This has also contributed to the extrapolation of crystallization kinetics. XRD is often combined with other techniques such as DSC and polarized light microscopy for full assessment of a single component or an additive on crystallization of the whole sample (Verstringe et al., 2013). Enhancement of traditional XRD with synchrotron (around which high-energy electrons travel and release X-rays) has improved the performance of this device, higher resolution and time resolution increasing the efficiency of the method. It has been used to assess a wide array of products. In a study by Mazzanti et al. (2009), crystallization kinetics of anhydrous milk fat and milk fat triglycerides were studied at different temperatures (17,17.5, and 20°C) coupled with shear stress (0–2880 s^{-1}). It was observed that onset time for phase α was not affected by the shear; however, the shear effect was observed during the transition from α to β' phase. The growth of β' phase and its division into two stages was also studied; "series parallel system of differential equations" were employed for the description of first phase of conversion from liquid to crystalline phases and diffusion was predominant in the second stage. In a different study by Zakharov et al. (2016) on the effect of a generic protein on ice recrystallization, synchrotron XRD was also employed as a unique method for the analysis and quantification of crystallinity in samples, and disorder in frozen aqueous solutions. Synchrotron XRD has also been combined with DSC for further enhanced characterization of the crystallization process mechanisms occurring during temperature changes and their correlation with the corresponding thermal processes taking place during the melting and crystallization processes of extra virgin olive oil (Barba et al., 2013). Cai et al. (2012) studied the melting and crystallization of α 1,4-glucan (short linear) obtained from various sources of debranched waxy starch, including potato and maize, by in-situ synchrotron wide-angle XRD. The differences in crystallinity implied that short-chain amylopectin crystallized more rapidly than longer chains; however, the corresponding crystals were found to be weaker. In another different study, Raman spectroscopy was combined with XRD to understand the polymorphic phase transitions occurring in different solvents during crystallization. Real-time measurements were taken that allowed the crystal content to be determined from formation of the first crystalline molecular aggregates to the final stable crystal form (Klimakow et al., 2010). Time-resolved synchrotron XRD was combined with DSC to identify the crystallization properties and polymorphism of triglycerides in fat globules of milk (Lopez et al., 2007). In this study, the milk fat phase demonstrated a complex polymorphism and crystals of six different types were identified. Some of these crystals also coexisted, and their progression in terms of time and temperature was quantitatively supervised. This process also helped determine the various crystallization behaviors observed in the various sizes of milk fat globules found during the production of different dairy products, with tempering periods in between unit operations. Hence, generic XRD applications have contributed to enhanced study of the crystallization mechanism of different food components, which has been further enhanced by application of the corresponding combined and advanced versions of the technique.

X-ray and neutron scattering are appropriate for the analysis of materials at a length scale other than those considered during crystallography and microscopy with spatial resolution between these groups of techniques. In the case of analyses of starches, "quantification of the lamellar architecture of semi crystalline growth rings" in the basic starch crystals has been described as the most significant advantage (Blazek and Gilbert, 2011). Small-angle X-ray scattering was combined with high-throughput microfluidics for observation of the protein crystallization occurring from the solution, and it was concluded that this is a potential technique for the analyses of protein interactions aimed at preparing corresponding phase diagrams for the demonstration of the crystallization mechanism. During the respective studies by Sadeghpour et al. (2018) and Ladd Parada et al. (2018), metastable polymorphs of triacylglycerols were quantified and characterized using small-angle X-ray scattering. Electron density profiles collected from the scattering method data were extrapolated using global fitting analysis. In addition to the metastable polymorphs, tilt angles of fatty acids, lamellar thickness, and the molten fraction of triacylglycerols were measured, which together demonstrated the early stages of crystal growth. This method has also evolved as an in-situ technique that can provide a better understanding of nucleation and growth during the metastable stages.

Other methods such as ultrasound have also been effective in characterizing crystallization and have helped with kinetic studies. In the crystallization of pork fat during cold storage, ultrasonic measurements were useful for characterizing the textural changes to the pork fat that occurred during the two-step process due to the separate crystallization of different compositional triacylglycerols (Corona et al., 2014). The ultrasonic shear reflection technique has been used to measure, interpret, or monitor the crystallization process in cocoa butter, where shear wave propagation was initially analyzed theoretically using liquid, gas, or solid-layer configurations with different thicknesses for calculation and interpretation of shear wave reflection coefficient equations; these equations were further linked with the shear wave reflection coefficient during cocoa-butter crystallization (Rigolle et al., 2015).

7.5.2 Processing Techniques

Pre-treatments and processing methods can modulate product properties, leading to better products. Crystallization processes generally consist of lowering the temperature below melting point or concentrating the solution above solubility concentration of the solute. Factors such as temperature, humidity, rate of cooling, freezer air-flow rate, and freeze–thaw cycles are also relevant (Payne et al., 2019). In certain cases, where high-speed agitation has been employed to achieve enhanced crystallization of fat crystals and the targeted product properties, increased crystal surface area, increased solid fat content and decreased elasticity have been observed (Nii and Takayanagi, 2014; Silva et al., 2017). Other methods employed to enhance the generic crystallization process to obtain a final crystallized product include application of ultrasound, modern drying methods, application of electric and magnetic fields and/or anti-solvents. Chemical compositions are also modified in different cases to amend and/or control the crystallization mechanisms or crystal properties (size, shape, etc.), which are further combined with the various advanced techniques mentioned above to obtain the targeted product. Other process parameters, such as pH of the solution, presence of particular components, effects of pressure, etc., are also important factors contributing to crystallization behavior and mechanism (Sánchez-García et al., 2019; Zulkurnain et al., 2016).

Microstreaming and cavitation occurring during high-frequency sonication can contribute towards improving and controlling the process through enhanced nucleation and modification of polymorphism (Payne et al., 2019). However, during this process, undesirable oxidation and off flavors may occur due to the free radical production caused by the higher frequency of ultrasound (Payne et al., 2019). Ultrasonication has also been employed for the separation of lactose from whey (Sánchez-García et al., 2018a) and for the enhancement of crystallization of calcium phosphate from whey (Kezia et al., 2016). During a study on production of lard emulsion with reduced cholesterol levels, samples prepared with ultrasound application along with whey protein isolates had smaller crystal size with higher uniform distribution, but slightly destabilized emulsions. The study concluded that application of ultrasound changed the fat crystalline structure and the metastability of the emulsion, and these could be valuable ingredients of emulsion-based healthy food products (Jiang et al., 2019). In a study by Singh et al. (2019), high-intensity ultrasound at 20 kHz (750 W) was applied during the crystallization of sucrose, following the addition of crystal seed, which led to removal of the latent period and enhanced the rate of crystallization (crystallization kinetics followed the first order). The latent period is defined as the initial period of crystallization, when the crystallization rate decreases, causing incomplete utilization of the sucrose from the solution (Singh et al., 2019). A detailed account of ultrasonic crystallization with emphasis on lactose crystallization is given by Zamanipoor and Mancera (2014). Many recent research studies have focused on the use of this method for the improvement of the crystallization mechanism in milk fat and plant oil-based blends, mainly in order to enhance nutritional value. In the case of anhydrous milk fat and rapeseed oil blends, it was observed that the application of ultrasound accelerated the crystallization process and reduced crystal size. Furthermore, melting behavior and secondary oxidation products were not affected by sonication (Gregersen et al., 2019). In a different study by Wagh et al. (2013), application of high-intensity ultrasonication also enhanced the crystallization of anhydrous milk fat fortified with lactose monolaurate in aqueous solution. In anhydrous milk fats, sonication led to creation of spherulite structures of uniform sizes. The corresponding macroscopic properties were modified due to nano and microscopic changes (Frydenberg et al., 2013). In the case of application of ultrasonication to blends with beeswax in different edible oils, crystallization of beeswax was enhanced and phase separation of the oils was delayed (Jana and Martini, 2014).

Application of electric and magnetic fields causes increased nucleation of ice crystals through alignment of water molecules. Often freezing is also assisted by other advanced electrical or magnetic field applications. Insights on the combination of freezing and other methods have been summarized by Dalvi-Isfahan et al. (2017). Electro-freezing can be categorized into two types according to the mode of application of the electric field: either through use of a charged surface or by application of external electric fields. In the case of nucleation through an electrostatically charged surface, continuous or pulsed electric fields are passed through closely spaced electrodes, which are generally immersed in solutions; examples include preparation of pharmaceuticals or products with larger ice crystals (and pores) by freeze-drying at higher nucleation temperatures. In the case of external electric field applications, high-voltage DC is applied (with no voltage or current fluctuations) to a sample placed between electrodes during freezing. The electro-freezing mechanism is not yet comprehensively understood; however, the major principles of electro-freezing through static electric fields (based on molecular dynamic simulations) include interaction of the fields and polarization of water molecule dipoles by the electrostatic field. Thermodynamically, it has been deduced that the energy barrier for phase transition is reduced by the applied electric field, allowing nucleation of water through a lower degree of super cooling. Electromagnetic radiation (including microwave and radiofrequency) applications have also led to promotion of ice nucleation and decreased crystal size, as observed in the microstructure of meat with a better cellular network and an increased number of nucleation sites (Xanthakis et al., 2014). Different aspects of the application of magnetic fields to freezing have been briefly summarized by Dalvi-Isfahan et al. (2017). Jha et al. (2017a) have also reviewed in detail the effect of electric, magnetic, and electromagnetic fields on food matrices and the formation of ice crystals.

Anti-freeze proteins are added as an ice-structuring protein to different products including meat products, ice cream, and other frozen desserts. These proteins contribute to the mechanism and control of nucleation, crystal growth and changes occurring during storage. The process of crystallization using anti-freezing agents can have a huge impact on super-chilling. Generally, anti-freeze proteins interact with the ice surface, affecting ice growth, and these proteins contribute to lowering the freezing point of water without affecting its melting point through adsorption inhibition. Thus, these biomolecules can retard the growth of ice crystals and recrystallization in the biomatrices, and the consequent cellular damage in frozen food products (Kaale and Eikevik, 2014; Payne et al., 2019).

Crystallization using anti-solvent causes changes in the properties of the solution and solubility of the targeted crystalline component (in semi-continuous isothermal mode), where organic solvents have been more prevalent, for example, a combination of ethanol, DMSO and water for the crystallization of αα-lactose-monohydrate (Parimaladevi and Srinivasan, 2016) and acetonitrile for D-sorbitol crystallization (Hu et al., 2017). However, water has also been used as an anti-solvent for monoacylglycerols (Verstringe et al., 2015). For anti-solvent crystallization, the corresponding anti-solvent is generally added (at constant or discontinuous flow rate) to a solution saturated with the crystalline component, kept at the same temperature as the anti-solvent. This technique generates a high yield (with lower energy costs) and has been used for thermo sensitive components. However, the flow of anti-solvent should be properly controlled, as higher or lower flow rates could lead to extremely small crystals susceptible to agglomeration. The study of this process includes the properties and flow rate of the anti-solvent affecting the process, and the effect of other processing/operating parameters on the metastable zone width (El Bazi et al., 2018). Ethanol has frequently been used for the crystallization of glycine from the corresponding aqueous solution and lactose from whey (El Bazi et al., 2018; Gajendragadkar and Gogate, 2017), further enhanced by application of ultrasonic irradiation, as reported in several other studies. In the case of crystallization of glycine, the addition of ethanol caused a decrease in solubility and widened the metastable zone; however, it did not have a noticeable effect on the crystal size distribution (El Bazi et al., 2018). In a different study by Nii and Takayanagi (2014), when anti-solvent crystallization was performed under 1.6 MHz ultrasonic irradiation, the growth of α-glycine crystal and crystal size consistency increased. Lactose, the major component of whey, is recovered from whey through anti-solvent crystallization, a process that has been further enhanced in both quality and quantity through sono-crystallization for an optimum time interval and with optimum frequency and power levels (Gajendragadkar and Gogate, 2017).

Thus, crystallization methods can be employed sequentially or in combination to enhance the crystallization mechanism and crystal properties. The combined application of ultrasound and anti-solvents has been observed as an efficient technology for lactose crystallization, which can be used for the optimization of lactose properties relevant to pharmaceuticals manufacture (Kougoulos et al., 2010). Simultaneous addition of food additives and the application of ultrasound can also modify the generic crystallization mechanism of different biochemicals. The addition of whey proteins and carrageenan to aqueous lactose solutions accelerated the crystallization process, generated smaller crystals with the narrowest crystal size distribution, and decreased formation of amorphous lactose (Sánchez-García et al., 2018b). Ultrasound-assisted freezing has been employed for various products including ice cream and for fruits and vegetables. In the case of ice cream, targeted small and uniform ice crystals were formed efficiently (Russell et al., 1999) and in the case of fruits and vegetables, this combination technique led to an increased freezing rate and numerous small crystals without any deterioration of the generic cellular structure (Lili et al., 2013; Zhang et al., 2015). In a study by Ike and Hirasawa (2017), it was observed that L-ArgHCl exhibited polymorphism with three different crystal forms. This was overcome by application of ultrasonic radiation after the addition of an anti-solvent, and the process reduced induction time.

Freeze-drying, a popular technique for the preservation of food commodities and food flavors, is affected by the crystallization mechanism, the microstructure and other physico-chemical properties and mechanisms observed during pre-freezing and freeze-drying (Harnkarnsujarit et al., 2016; Zhao et al., 2016). During freeze-drying, as the moisture evaporates, the porosity of the product increases, and the remaining solid network is expected to take on the structural attributes of a porous structure. However, when the temperature of the product is higher than the glass transition temperature, its viscosity might not be sufficient to support the porous structure, causing it to collapse and the final product to shrink. Products with high sugar content possess lower collapse temperatures; hence, the drying of fruits rich in sugar has traditionally caused shrinkage. The crystalline state is the most stable state of sugar, attributed to the lowest free energy level. During the freeze-drying of pequi fruit, it was observed that individual pre-treatment of the fruit with ethanol and sugar led to amorphous structures, and freeze-drying of samples along with pre-treatment with a combination of ethanol and sugar led to crystallized sugar structures, induced during the freeze-drying process (de Oliveira Alves et al., 2010). The major components undergoing crystallization processes in these cases are generally sugars and water.

7.5.3 Advanced Computational Methods for Crystallization

Advanced analytical methods have augmented the characterization of crystal properties and the crystallization mechanism, and this has been further enhanced through the application of computational techniques for optimization of the process parameters and study of the crystal properties.

Study and analysis of different factors in crystallization and the corresponding properties of the product can be performed using regression modeling of a standard or advanced experimental design. Crystallization phenomena observed in complex formulations with different ingredients such as fat and sugar are very different from bulk systems; bulk-system observations cannot be extrapolated to predict crystallization behavior in a complex matrix. West and Rousseau (2019) analyzed and modeled the effect of processing temperature and confectioner's sugar on the crystallization behavior and the corresponding rheological properties of palm oil, using multivariate regression analysis with a binary variable, to create predictive models that corrected the effects of confectioner's sugar, including fat crystal morphology and increased network rigidity. Molecular dynamics simulations have been reported as useful technique for precise measurement of the "conformational distribution" of molecules in solution, which in turn is helpful for interpreting the energy required by individual molecules for movement from solution (fluid) to the crystalline state. This information can assist understanding of the controlling mechanism for the acceleration or retardation of the corresponding process to achieve the targeted properties in a commodity (Payne et al., 2019). Recently, deep neural networks have also been used to monitor sugar crystallization, which can be utilized for automation control, leading to a reduction in the labor force. In this study, a deep-learning framework was built based on "deep convolutional neural networks" to develop a system for accurate image classification of sugar crystals. The large data set collected during the study was used to train different networks and was analyzed to build a model with classification accuracy of 0.901, which was further used to differentiate and classify these images (Zhang et al., 2020c). Naik et al. (2019) combined molecular dynamics simulations and artificial neural networks to analyze the crystallization process in Indian honey. Sugar contents and "moisture content-based mixture systems" of different honeys were used for molecular dynamics simulations. All the simulations led to stable crystals, and the combination of fructose and glucose (1.18) was found to be the most stable crystal with the highest electrostatic and vander Waals interactions. Using the simulation data obtained during the study, an artificial neural network was trained to predict the stability of honey crystallization in terms of glucose:water and fructose:glucose ratios. Wong et al. (2010) used an artificial neural network to model the crystallization kinetics rate of lactose. Different research groups have also built or used structural analysis algorithms and predictive models with optimization algorithms for controlling the process and observing the structural differences during crystallization (Doesburg and Nierop, 2013; Jha et al., 2017b). Advanced sensors, scanners, and analytical tools such as artificial neural network applications can further elucidate crystallization at industrial scale (Damour et al., 2010a, 2010b).

7.6 FUTURE PROSPECTS AND RESEARCH

A number of other methods of characterizing and controlling the crystallization mechanism have been used in recent years, though not as extensively as others. During a study of the effect of polyglycerol esters addition on palm oil crystallization, a focused beam reflectance measurement method was used to study in-line deviations in the crystal size distribution (Saw et al., 2017).

Many studies have emphasized the effect of crystallization technique parameters on different properties of products (Miyagawa et al., 2016; Payne et al., 2019). An increasing number of studies not only focus on the characterization of crystal properties or the crystallization mechanism, but also highlight the effect of crystals (quantity, physical, and surface properties) on various other functional properties. In a study by Goulart and Hartel (2017) on lactose crystallization in a milk protein concentrate, the differences in viscosities with total solid content were correlated with the content of crystal mass and not with the total solid content itself.

Development of diverse food products has also impacted research focused on their characterization techniques, their varying stability with different crystallization parameters, and the effect of different components. Recently, reduction of waste and research on the creation of value-added products have also been strongly emphasized. Techniques for characterization and control of the crystallization process have been further optimized, depending on the product of interest. Nanocrystals obtained/extracted from various sources are being exploited; hence, properties of these nanocrystals have been characterized ahead of further utilization. Cellulose nanocrystals are generally prepared using acid digestion methods, which also find applications in medicine, as a bio-composite, in food packaging materials, and in water treatment. Scanning electron microscopy (for the characterization of crystal shape, surface topography, and composition), XRD and NMR spectroscopy are some of the advanced technologies used for characterization (Abiaziem et al., 2019).

7.6.1 Effect of Research Advances in Novel Product Development

Crystallization is extensively affected by the change in composition of traditional commodities and the inclusion of additives. Hence, in the case of emulsions, different aspects of surfactant crystallization are also being analyzed. Goibier et al. (2017) studied partial coalescence in oil/water emulsions and the effect of surfactant crystallization. In emulsions where the dispersed phase is incompletely crystallized, instability resulting in partial coalescence was always observed. When stabilizing proteins were added, instability was hindered due to repulsive surface forces and interfacial stiffness. Low molecular weight surfactants have been shown to promote partial coalescence; however, when crystallizable surfactants (especially liquid ones) are added, marginal partial coalescence is observed, further affecting product properties.

A vast array of research also concentrates on the formation of crystallized bioactive components. Crystallization of bioactive extracts with other components such as *Basellarubra extract* containing betacyclin with sucrose (Karangutkar and Ananthanarayan, 2019), succinic acid extracted from *Actinobacillus succinogenes* ATCC 55618 with fresh cassava root (Thuy et al., 2017), milk fat stearin with omega-3 rich oils (Li et al., 2017), and formation of bioactive crystalline forms such as crystalline quercetin hydrate (Klitou et al., 2019) are some of the topics that have been recently investigated. Health awareness has further promoted research into functional biomolecules such as resistant starch components. Resistant starch formation through crystallization of starchy materials is achieved through the re-association of amylose chains and the formation of double helices which loosely arrange to form a partially crystalline system resisting hydrolytic enzyme binding and diffusion. Crystallization mechanisms and behaviors, and the formation of these resistant starches, are affected by concentration, temperature, and chain distribution, as well as by hydrothermal treatment, which has been shown to improve resistant starch formation. However, the efficiency of the process also depends on the perfection of the initial crystalline form (Kiatponglarp et al., 2015). Consumption of slowly digestible starch has also recently been encouraged, because of the slow increase postprandial blood glucose level, maintenance of blood glucose levels, and the consequent prevention of hyperglycemia-related diseases. Higher relative crystallinity has been observed with storage under temperature-cycled conditions, which have been shown increases the amount of slowly digestible starches compared to isothermal storage (Zeng et al., 2014).

Several vegetable oil combinations are also used as substitutes for trans-fat and/or cocoa butter; hence their crystallization mechanisms are analyzed to observe the different polymorphic forms obtained with different methods and temperatures. Isothermal crystallization of two high-stearic high-oleic sunflower oil stearins was conducted by Rincón-Cardona et al. (2013), who obtained three polymorphs. The β_2 polymorphic form required for chocolate manufacturing appeared for a very short time; hence, it was isolated from β_1 employing heating/cooling cycles. When new sources of food fortification are being used, process parameters should be efficiently controlled to develop the desired product quality.

Research projects at lab scale and pilot scale focused on commercial fortification of enzymes and other nutraceutical components have increased in recent decades (Pereira-Rodriguez et al., 2010), which may result in further diversification of crystallization studies according to raw material, base food matrix, bioavailability of the component, and targeted consumer groups. Some of the products that have been recently investigated are summarized in Table 7.2.

Increasing concern about environmental pollution from the overuse of plastics in different sectors has also augmented research into development of highly functional biodegradable bioplastics, applicable in different sectors. Polylactic acid-based

TABLE 7.2
Novel Food Products Based on the Crystallization Principle

Product	Crystal Components	Main Findings	References
Lard emulsion with reduced cholesterol	Fat crystals	Ultrasound treatment along with addition of protein emulsifiers	Jiang et al. (2019)
Anhydrous milk fat and rapeseed oil blends	Fat crystals	Sonication-accelerated crystallization, especially in blends with higher rapeseed oil content and smaller crystal size	Gregersen et al. (2019)
Edible oleogels to be used as trans-fat and saturated fat replacers	Beewax in six different oils	Higher wax content leads to faster crystallization along with higher turbidity. Ultrasonication prevents phase separation and improves crystallization of beeswax.	Jana and Martini (2014)
Anhydrous milk fat solution fortified with lactose monolaurate	Milk fat	High-intensity ultrasound application leads to enhanced crystallization of anhydrous milk fat	Wagh et al. (2013)
Anhydrous milk fat-sunflower oil wax blends	Milk fat	Addition of waxes induces start of crystallization	Kerr et al. (2011)
Coconut and palm olein with monoacylglycerols emulsifiers.	Oil	High-intensity ultrasound induced crystallization in most of the samples. Effect of ultrasonication and addition of emulsifiers on the crystallization mechanism and efficiency of these oils observed.	Maruyama et al. (2016)
Polyglycerol esters addition in palm oil	Fat crystal	Polyglycerol esters enhance nucleation of fat crystals but retard growth rate of crystals	Saw et al. (2017)
Mixtures of diacylglycerol-enriched palm oil blends with diacylglycerol-enriched palm olein	Fat crystal	Blending leads to higher crystallization rates with no spherulitic growth	Xu et al. (2016b)
Recombined milk cream with monoacylglycerols	Milk fat crystals	Interfacial heterogeneous nucleation induced by long chain saturated monoacylglycerols along with enhancement of crystal growth and α-β' polymorphic transition	Fredrick et al. (2013)
Berry wax and sunflower wax fortified rice bran oil	Wax crystals	—	Doan et al. (2017)
β-galactosidase from *Kluyveromyceslactis*	β-galactosidase	Further study required. Crystals obtained at optimum parameters possess poor diffraction quality, which improves with streak seeding.	Pereira-Rodriguez et al. (2010)
Polylactic acid-based bioplastics with cellulose nanofibers and nanocrystals	Polylactic acid	Cellulose nanocrystals enhance crystallization and contribute to nucleation of polylactic acid.	Clarkson et al. (2019)
Vitamins in amorphous sugar systems	Sugars	Crystallization of lactose-enhanced vitamin degradation. Trehalose retains much higher vitamin concentration.	Zhou and Roos (2012)

products reinforced with nanocrystals are being used as a raw material for food packaging. Clarkson et al. (2019) observed that cellulose nanocrystals and cellulose nanofibers acted as efficient nucleating agents, and crystallization was enhanced in the presence of cellulose nanocrystals even at lower concentrations. Further research on the development of similar products will also require extensive study of the effect of components and parameters on the crystallization behavior of the respective matrices.

Extraction, purification, characterization, and utilization of nanocrystals have recently initiated a huge new area of multi-faceted research, which has further enhanced research in associated areas. Some of the nanocrystals that have attracted interest from research groups worldwide include cellulose crystals from food industry by-products, such as lemon seed (Zhang et al., 2020b), starch nanocrystals (Zhang et al., 2020a), chitin and chitosan nanocrystals (Hadidi et al., 2020; Yuan et al., 2020), and metal nanocrystals (Affes et al., 2020; Araruna et al., 2019). Cellulose and starch nanocrystals have been mainly employed as structural elements in different biodegradable composite materials (Clarkson et al., 2019; Zhang et al., 2020a) and gels (Xiao et al., 2020). Other applications of nanocrystals include use as a component of bioactive anti-bacterial and antioxidant supplements (Affes et al., 2020; Hadidi et al., 2020). Major extraction methods include solvent extraction (with acids, deep eutectic solvents) (Yuan et al., 2020; Zhang et al., 2020b) and microwave synthesis (Araruna et al., 2019).

7.7 CONCLUSIONS

Advanced analytical technologies and processing techniques have greatly increased understanding of crystallization principles and mechanisms. Additional ongoing and future research in this regard will further enhance knowledge of the principles and mechanisms of crystallization of complex biomolecules such as fat nanocrystals, and may lead to measures completely controlling ice recrystallization in frozen products. Advances in crystallization study will also encourage development of novel products including bioavailable nanocrystal-based pharmaceuticals, nutraceuticals, and food products, positively impacting consumer health. Understanding of the principles and mechanisms of crystallization for the formation of enhanced biomatrices can further contribute to the manufacture of active packaging materials. Crystallization is an evolving subject with impacts in all aspects and segments of commodities and the food industry, and is thus deserving of further exploration.

ACKNOWLEDGMENTS

The author would like to thank Dr. R.C. Pradhan (Head of the Department of Food Process Engineering, NIT Rourkela) for continual support during the project.

REFERENCES

Abd. Rashid, N., C. Chiew Let, C. Chong Seng, and Z. Omar. 2012. Crystallisation kinetics of palm stearin, palm kernel olein and their blends. *LWT - Food Science and Technology*, 46(2): 571–573.

Abiaziem, C.V., A.B. Williams, A.I. Inegbenebor, C.T. Onwordi, C.O. Ehi-Eromosele, and L.F. Petrik. 2019. *Preparation and characterisation of cellulose nanocrystal from sugarcane peels by XRD, SEM and CP/MAS 13C NMR. Paper read at Journal of Physics: Conference Series.*

Acevedo, N.C., and A.G. Marangoni. 2010. Characterization of the nanoscale in triacylglycerol crystal networks. *Crystal Growth & Design.* 10(8): 3327–3333.

Acevedo, N.C., and A.G. Marangoni. 2015. Nanostructured fat crystal systems. *Annual Review of Food Science and Technology*, 6: 71–96.

Adhikari, B.M., T. Truong, N. Bansal, and B. Bhandari. 2018. Influence of gas addition on crystallisation behaviour of lactose from supersaturated solution. *Food and Bioproducts Processing*, 109: 86–97.

Affes, S., H. Maalej, I. Aranaz, H. Kchaou, N. Acosta, Á. Heras, and M. Nasri. 2020. Controlled size green synthesis of bioactive silver nanoparticles assisted by chitosan and its derivatives and their application in biofilm preparation. *Carbohydrate Polymers.* 236: 116063.

Alvarez, L., P. Urrutia, A. Olivares, A. Flores, B. Bhandari, T. Truong, and S. Almonacid. 2020. Impact of thermal pretreatment on crystallization of Thompson raisins. *Food Chemistry*. doi: 10.1016/j.foodchem.2020.126381:126381.

Araruna, F.B., T.M. de Oliveira, P.V. Quelemes, A.R. de Araújo Nobre, A. Plácido, A.G. Vasconcelos, R.C.M. de Paula, A.C. Mafud, M.P. de Almeida, C. Delerue-Matos, Y.P. Mascarenhas, P. Eaton, J.R. de Souza de Almeida Leite, and D.A. da Silva. 2019. Antibacterial application of natural and carboxymethylated cashew gum-based silver nanoparticles produced by microwave-assisted synthesis. *Carbohydrate Polymers.* doi: 10.1016/j.carbpol.2019.115260:115260.

Barba, L., G. Arrighetti, and S. Calligaris. 2013. Crystallization and melting properties of extra virgin olive oil studied by synchrotron XRD and DSC. *European Journal of Lipid Science and Technology*, 115(3): 322–329.

Bayard, M., F. Leal-Calderon, and M. Cansell. 2017. Free fatty acids and their esters modulate isothermal crystallization of anhydrous milk fat. *Food Chemistry*, 218: 22–29.

Beck, M., M. Jekle, and T. Becker. 2011. Starch re-crystallization kinetics as a function of various cations. *Starch-Stärke*, 63(12): 792–800.

Beckmann, W. 2013. *Crystallization: Basic Concepts and Industrial Applications.* John Wiley & Sons.

Bensouissi, A., B. Roge, and M. Mathlouthi. 2010. Effect of conformation and water interactions of sucrose, maltitol, mannitol and xylitol on their metastable zone width and ease of nucleation. *Food Chemistry*, 122(2): 443–446.

Bhandari, B.R., and Y.H. Roos. 2016. *Non-Equilibrium States and Glass Transitions in Foods: Processing Effects and Product-Specific Implications.* Woodhead Publishing.

Blazek, J., and E.P. Gilbert. 2011. Application of small-angle X-ray and neutron scattering techniques to the characterisation of starch structure: A review. *Carbohydrate Polymers*, 85(2): 281–293.

Bund, R., and R. Hartel (eds.). 2010. Crystallization in foods and food quality deterioration. In: *Chemical Deterioration and Physical Instability of Food and Beverages.* Amsterdam: Elsevier, pp. 186–215. Original edition.

Buscato, M.H.M., B.G. Zaia, K.R.R.D. Godoi, A.P.B. Ribeiro, and T.G. Kieckbusch. 2018. Modification of palm oil crystallization by phytosterol addition as a tool for structuring a low saturated lipid blend. *Brazilian Journal of Chemical Engineering.* 35(1): 169–180.

Cai, L., Y. Bai, and Y.-C. Shi. 2012. Study on melting and crystallization of short-linear chains from debranched waxy starches by in situ synchrotron wide-angle X-ray diffraction. *Journal of Cereal Science*, 55(3): 373–379.

Chai, X., Z. Meng, J. Jiang, P. Cao, X. Liang, M. Piatko, S. Campbell, S.K. Lo, and Y. Liu. 2018a. Non-triglyceride components modulate the fat crystal network of palm kernel oil and coconut oil. *Food Research International*, 105: 423–431.

Chai, X.-H., Z. Meng, P.-R. Cao, X.-Y. Liang, M. Piatko, S. Campbell, S.K. Lo, and Y.-F. Liu. 2018b. Influence of indigenous minor components on fat crystal network of fully hydrogenated palm kernel oil and fully hydrogenated coconut oil. *Food Chemistry*, 255: 49–57.

Chaleepa, K., A. Szepes, and J. Ulrich. 2010. Effect of additives on isothermal crystallization kinetics and physical characteristics of coconut oil. *Chemistry and Physics of Lipids*, 163(4): 390–396.

Chen, J., C. Nowakowski, D. Green, and R.W. Hartel. 2015. State behavior and crystal growth kinetics of sucrose and corn syrup mixtures. *Journal of Food Engineering*, 161: 1–7.

Clarkson, C.M., S.M.E.A. Azrak, G.T. Schueneman, J.F. Snyder, and J.P. Youngblood. 2019. Crystallization kinetics and morphology of small concentrations of cellulose nanofibrils (CNFs) and cellulose nanocrystals (CNCs) melt-compounded into poly (lactic acid) (PLA) with plasticizer. *Polymer*. 122101.

Cook, K., and R. Hartel. 2010. Mechanisms of ice crystallization in ice cream production. *Comprehensive Reviews in Food Science and Food Safety*, 9(2): 213–222.

Corona, E., J.V. García-Pérez, J.V. Santacatalina, S. Ventanas, and J. Benedito. 2014. Ultrasonic characterization of pork fat crystallization during cold storage. *Journal of Food Science*. 79(5): E828–E838.

Crafts, P. 2007. The role of solubility modeling and crystallization in the design of active pharmaceutical ingredients, pp. 23–85 inK.M. Ng, R. Gani, and K. Dam-Johansen (eds.), *Computer Aided Chemical Engineering*. Amsterdam: Elsevier. Original edition.

Daels, E., A. Rigolle, K. Raes, J. De Block, and I. Foubert. 2015. Monoglycerides, polyglycerol esters, lecithin, and their mixtures influence the onset of non-isothermal fat crystallization in a concentration dependent manner. *European Journal of Lipid Science and Technology*, 117(11): 1745–1753.

Dalvi-Isfahan, M., N. Hamdami, E. Xanthakis, and A. Le-Bail. 2017. Review on the control of ice nucleation by ultrasound waves, electric and magnetic fields. *Journal of Food Engineering*, 195: 222–234.

Damour, C., M. Benne, B. Grondin-Perez, and J.-P. Chabriat. 2010a. Nonlinear predictive control based on artificial neural network model for industrial crystallization. *Journal of Food Engineering*, 99(2): 225–231.

Damour, C., M. Benne, B. Grondin-Perez, and J.-P. Chabriat. 2010b. Soft-sensor for industrial sugar crystallization: On-line mass of crystals, concentration and purity measurement. *Control Engineering Practice*, 18(8): 839–844.

Das, D., and T.A. Langrish. 2012. Activated-rate theory: Effect of protein inhibition and the temperature dependence of crystallization kinetics for lactose-protein mixtures. *Food Research International*, 48(2): 367–373.

de Oliveira Alves, C.C., J.V. de Resende, M.E.T. Prado, and R.S.R. Cruvinel. 2010. The effects of added sugars and alcohols on the induction of crystallization and the stability of the freeze-dried peki (Caryocar brasiliense Camb.) fruit pulps. *LWT - Food Science and Technology*, 43(6): 934–941.

de Oliveira, G.M., A.P. Badan Ribeiro, A.O. dos Santos, L.P. Cardoso, and T.G. Kieckbusch. 2015. Hard fats as additives in palm oil and its relationships to crystallization process and polymorphism. *LWT - Food Science and Technology*, 63(2): 1163–1170.

Dettori, A., S. Tappi, L. Piana, M. Dalla Rosa, and P. Rocculi. 2018. Kinetic of induced honey crystallization and related evolution of structural and physical properties. *LWT - Food Science and Technology* 95: 333–338.

Di Bari, V., W. Macnaughtan, J. Norton, A. Sullo, and I. Norton. 2017. Crystallisation in water-in-cocoa butter emulsions: Role of the dispersed phase on fat crystallisation and polymorphic transition. *Food Structure*, 12: 82–93.

Doan, C.D., I. Tavernier, M.D.B. Sintang, S. Danthine, D. Van de Walle, T. Rimaux, and K. Dewettinck. 2017. Crystallization and gelation behavior of low-and high melting waxes in rice bran oil: A case-study on berry wax and sunflower wax. *Food Biophysics*, 12(1): 97–108.

Doesburg, P., and A.F. Nierop. 2013. Development of a structure analysis algorithm on structures from $CuCl_2\cdot 2H_2O$ crystallization with agricultural products. *Computers and Electronics in Agriculture*, 90: 63–67.

El Bazi, W., M.-T. Moufarej Abou Jaoude, C. Porte, I. Mabille, and J.-L. Havet. 2018. Isothermal crystallization of glycine in semi-continuous mode by anti-solvent addition. *Journal of Crystal Growth*, 498: 202–208.

Erdemir, D., A.Y. Lee, and A.S. Myerson. 2009. Nucleation of crystals from solution: Classical and two-step models. *Accounts of Chemical Research*, 42(5): 621–629.

Fan, F., and Y.H. Roos. 2016. Crystallization and structural relaxation times in structural strength analysis of amorphous sugar/whey protein systems. *Food Hydrocolloids*, 60: 85–97.

Ferguson, S., G. Morris, H. Hao, M. Barrett, and B. Glennon. 2012. In-situ monitoring and characterization of plug flow crystallizers. *Chemical Engineering Science*, 77: 105–111.

Fredrick, E., K. Moens, B. Heyman, S. Fischer, P. Van der Meeren, and K. Dewettinck. 2013. Monoacylglycerols in dairy recombined cream: I. The effect on milk fat crystallization. *Food Research International*, 51(2): 892–898.

Frydenberg, R.P., M. Hammershøj, U. Andersen, and L. Wiking. 2013. Ultrasonication affects crystallization mechanisms and kinetics of anhydrous milk fat. *Crystal Growth & Design*, 13(12): 5375–5382.

Gajendragadkar, C.N., and P.R. Gogate. 2017. Ultrasound assisted intensified recovery of lactose from whey based on antisolvent crystallization. *Ultrasonics Sonochemistry*, 38: 754–765.

Garti, N., and K. Sato. 2001. *Crystallization Processes in Fats and Lipid Systems*. Boca Raton, FL: CRC Press.

Gharsallaoui, A., B. Rogé, and M. Mathlouthi. 2008. Water–disaccharides interactions in saturated solution and the crystallisation conditions. *Food Chemistry*, 106(4): 1329–1339.

Goibier, L., S. Lecomte, F. Leal-Calderon, and C. Faure. 2017. The effect of surfactant crystallization on partial coalescence in O/W emulsions. *Journal of Colloid and Interface Science*, 500: 304–314.

Goulart, D.B., and R.W. Hartel. 2017. Lactose crystallization in milk protein concentrate and its effects on rheology. *Journal of Food Engineering*, 212: 97–107.

Gregersen, S.B., R.P. Frydenberg, M. Hammershøj, T.K. Dalsgaard, U. Andersen, and L. Wiking. 2019. Application of high intensity ultrasound to accelerate crystallization of anhydrous milk fat and rapeseed oil blends. *European Journal of Lipid Science and Technology*, 121(1): 1800200.

Guo, E., G. Zeng, D. Kazantsev, P. Rockett, J. Bent, M. Kirkland, G. Van Dalen, D.S. Eastwood, D. StJohn, and P.D. Lee. 2017. Synchrotron X-ray tomographic quantification of microstructural evolution in ice cream–a multi-phase soft solid. *RSC Advances*, 7(25): 15561–15573.

Hadidi, M., S. Pouramin, F. Adinepour, S. Haghani, and S.M. Jafari. 2020. Chitosan nanoparticles loaded with clove essential oil: Characterization, antioxidant and antibacterial activities. *Carbohydrate Polymers*, 236: 116075.

Harnkarnsujarit, N., and S. Charoenrein. 2011. Effect of water activity on sugar crystallization and β-carotene stability of freeze-dried mango powder. *Journal of Food Engineering*, 105(4): 592–598.

Harnkarnsujarit, N., K. Kawai, M. Watanabe, and T. Suzuki. 2016. Effects of freezing on microstructure and rehydration properties of freeze-dried soybean curd. *Journal of Food Engineering*, 184: 10–20.

Harrison, P.D., K.W. Smith, K. Bhaggan, and A.G.F. Stapley. 2016. Image analysis of palm oil crystallisation as observed by hot stage microscopy. *Journal of Crystal Growth*, 444: 28–38.

Hartel, R.W. 2013. Advances in food crystallization. *Annual Review of Food Science and Technology*, 4: 277–292.

Hartel, R.W., H. Joachim, V. Elbe, and R. Hofberger. 2018. *Confectionery Science and Technology*. Springer.

Hu, Y., Y. Yu, Y. Zhang, W. Yang, T. Li, R. Deng, and M. Jiang. 2017. Experimental measurements and equilibrium study of functional d-sorbitol in good and anti-solvent binary mixtures. *Journal of Molecular Liquids*, 238: 296–302.

Hubbes, S.-S., W. Danzl, and P. Foerst. 2018. Crystallization kinetics of palm oil of different geographic origins and blends thereof by the application of the Avrami model. *LWT - Food Science and Technology*, 93: 189–196.

Ike, Y., and I. Hirasawa. 2017. Polymorph control of L-Arg HCl on antisolvent crystallization by ultrasonic irradiation. *Chemical Engineering & Technology*, 40(7): 1318–1322.

Jana, S., and S. Martini. 2014. Effect of high-intensity ultrasound and cooling rate on the crystallization behavior of beeswax in edible oils. *Journal of Agricultural and Food Chemistry*, 62(41): 10192–10202.

Jha, P.K., M. Sadot, S.A. Vino, V. Jury, S. Curet-Ploquin, O. Rouaud, M. Havet, and A. Le-Bail. 2017a. A review on effect of DC voltage on crystallization process in food systems. *Innovative Food Science & Emerging Technologies*, 42: 204–219.

Jha, S.K., S. Karthika, and T.K. Radhakrishnan. 2017b. Modelling and control of crystallization process. *Resource-Efficient Technologies*, 3(1): 94–100.

Jiang, J., Z. Song, Q. Wang, X. Xu, Y. Liu, and Y.L. Xiong. 2019. Ultrasound-mediated interfacial protein adsorption and fat crystallization in cholesterol-reduced lard emulsion. *Ultrasonics Sonochemistry*, 58: 104641.

Jouppila, K., J. Kansikas, and Y. Roos. 1997. Glass transition, water plasticization, and lactose crystallization in skim milk powder. *Journal of Dairy Science*, 80(12): 3152–3160.

Kaale, L.D., and T.M. Eikevik. 2014. The development of ice crystals in food products during the superchilling process and following storage, a review. *Trends in Food Science &Technology*, 39(2): 91–103.

Kadam, S.S., E. van der Windt, P.J. Daudey, and H.J. Kramer. 2010. A comparative study of ATR-FTIR and FT-NIR spectroscopy for in-situ concentration monitoring during batch cooling crystallization processes. *Crystal Growth & Design*, 10(6): 2629–2640.

Karangutkar, A., and L. Ananthanarayan. 2019. Co-crystallization of Basella rubra extract with sucrose: Characterization of co-crystals and evaluating the storage stability of betacyanin pigments. *Journal of Food Engineering*. 109776.

Kerr, R.M., X. Tombokan, S. Ghosh, and S. Martini. 2011. Crystallization behavior of anhydrous milk fat–sunflower oil wax blends. *Journal of Agricultural and Food Chemistry*, 59(6): 2689–2695.

Kezia, K., J. Lee, B. Zisu, M. Weeks, G. Chen, S. Gras, and S. Kentish. 2016. Crystallisation of minerals from concentrated saline dairy effluent. *Water Research*, 101: 300–308.

Kiani, H., and D.-W. Sun. 2011. Water crystallization and its importance to freezing of foods: A review. *Trends in Food Science & Technology*, 22(8): 407–426.

Kiatponglarp, W., S. Tongta, A. Rolland-Sabaté, and A. Buléon. 2015. Crystallization and chain reorganization of debranched rice starches in relation to resistant starch formation. *Carbohydrate Polymers*, 122: 108–114.

Klimakow, M., J. Leiterer, J. Kneipp, E. Rössler, U. Panne, K. Rademann, and F. Emmerling. 2010. Combined synchrotron XRD/raman measurements: In situ identification of polymorphic transitions during crystallization processes. *Langmuir*, 26(13): 11233–11237.

Klitou, P., I. Rosbottom, and E. Simone. 2019. Synthonic modelling of quercetin and its hydrates: Explaining crystallization behaviour in terms of molecular conformation and crystal packing. *Crystal Growth & Design*. doi: 10.1021/acs.cgd.9b00650.

Kougoulos, E., I. Marziano, and P.R. Miller. 2010. Lactose particle engineering: Influence of ultrasound and anti-solvent on crystal habit and particle size. *Journal of Crystal Growth*, 312(23): 3509–3520.

Ladd Parada, M., A. Sadeghpour, J. Vieira, M. Povey, and M. Rappolt. 2018. Global small-angle X-ray scattering data analysis of triacylglycerols in the α-phase (part II). *The Journal of Physical Chemistry B*, 122(45): 10330–10336.

Li, B.-Z., T. Truong, and B. Bhandari. 2017. Crystallization and melting properties of mixtures of milk fat stearin and omega-3 rich oils. *Food Chemistry*, 218: 199–206.

Li, S., Z. Xiong, P. Fei, J. Cai, H. Xiong, J. Tan, and Y. Yu. 2013. Parameters characterizing the kinetics of the nonisothermal crystallization of thermoplastic starch/poly (lactic acid) composites as determined by differential scanning calorimetry. *Journal of Applied Polymer Science*, 129(6): 3566–3573.

Lili, W., C. Yeming, and L. Zaigui. 2013. The effects of freezing on soybean microstructure and qualities of soymilk. *Journal of Food Engineering*, 116(1): 1–6.

Liu, C., Z. Zheng, Z. Meng, X. Chai, C. Cao, and Y. Liu. 2019. Beeswax and carnauba wax modulate the crystallization behavior of palm kernel stearin. *LWT - Food Science and Technology*, 108446.

Lopez, C., C. Bourgaux, P. Lesieur, and M. Ollivon. 2007. Coupling of time-resolved synchrotron X-ray diffraction and DSC to elucidate the crystallisation properties and polymorphism of triglycerides in milk fat globules. *Lait*, 87(4–5): 459–480.

Maruyama, J.M., A. Wagh, L.A. Gioielli, R.C.D. Silva, and S. Martini. 2016. Effects of high intensity ultrasound and emulsifiers on crystallization behavior of coconut oil and palm olein. *Food Research International*, 86: 54–63.

Mathew Thomas, K., and R. Lakerveld. 2019. An airlift crystallizer for protein crystallization. *Industrial & Engineering Chemistry Research*. doi: 10.1021/acs.iecr.9b03535.

Mazzanti, G., A.G. Marangoni, and S.H.J. Idziak. 2009. Synchrotron study on crystallization kinetics of milk fat under shear flow. *Food Research International*, 42(5): 682–694.

Meng, Z., L. Yang, W. Geng, Y. Yao, X. Wang, and Y. Liu. 2014. Kinetic study on the isothermal and nonisothermal crystallization of monoglyceride organogels. *The Scientific World Journal*. doi: 10.1155/2014/149753.

Mersmann, A. 2001. *Crystallization Technology Handbook*. Boca Raton, FL: CRC Press.

Miyagawa, Y., T. Ogawa, K. Nakagawa, and S. Adachi. 2016. Destabilization of mayonnaise induced by lipid crystallization upon freezing. *Bioscience, Biotechnology, and Biochemistry*, 80(4): 786–790.

Mousavi, R., T. Miri, P.W. Cox, and P.J. Fryer. 2007. Imaging food freezing using X-ray microtomography. *International Journal of Food Science & Technology*, 42(6): 714–727.

Mullin, J.W. 2001. *Crystallization*. Amsterdam: Elsevier.

Myerson, A., D. Erdemir, and A.Y. Lee. 2019. *Handbook of Industrial Crystallization*. Butterworth-Heinemann.

Naik, R.R., N.S. Gandhi, M. Thakur, and V. Nanda. 2019. Analysis of crystallization phenomenon in Indian honey using molecular dynamics simulations and artificial neural network. *Food Chemistry*, 300: 125182.

Nii, S., and S. Takayanagi. 2014. Growth and size control in anti-solvent crystallization of glycine with high frequency ultrasound. *Ultrasonics Sonochemistry*, 21(3): 1182–1186.

Palmieri, P.A., and R.W. Hartel. 2019. Crystallization of cocoa butter in cocoa powder. *Journal of the American Oil Chemists' Society*. 96(8): 911–926.

Parimaladevi, P., and K. Srinivasan. 2016. Understanding the crystallization behavior of α-lactose monohydrate (α-LM) through molecular interaction in selected solvents and solvent mixtures under different growth conditions. *Cryst Eng Comm*, 18(13): 2312–2318.

Payne, K., A. DeJong, and R.W. Hartel. 2019. Advances in crystallization in foods. In: *Reference Module in Food Science*. Amsterdam: Elsevier. Original edition.

Pereira-Rodriguez, A., R. Fernández-Leiro, M.I. Gonzalez Siso, M.E. Cerdán, M. Becerra, and J. Sanz-Aparicio. 2010. Crystallization and preliminary X-ray crystallographic analysis of β-galactosidase from Kluyveromyces lactis. *Acta Crystallographica Section F: Structural Biology and Crystallization Communications*, 66(3): 297–300.

Reinke, S.K., F. Wilde, S. Kozhar, F. Beckmann, J. Vieira, S. Heinrich, and S. Palzer. 2016. Synchrotron X-ray microtomography reveals interior microstructure of multicomponent food materials such as chocolate. *Journal of Food Engineering*, 174: 37–46.

Rigolle, A., I. Foubert, J. Hettler, E. Verboven, R. Demuynck, and K. Van Den Abeele. 2015. Development of an ultrasonic shear reflection technique to monitor the crystallization of cocoa butter. *Food Research International*, 75: 115–122.

Rincón-Cardona, J.A., S. Martini, R.J. Candal, and M.L. Herrera. 2013. Polymorphic behavior during isothermal crystallization of high stearic high oleic sunflower oil stearins. *Food Research International*, 51(1): 86–97.

Rizzo, G., J.E. Norton, and I.T. Norton. 2015. Emulsifier effects on fat crystallisation. *Food Structure*, 4: 27–33.

Russell, A., P. Cheney, and S. Wantling. 1999. Influence of freezing conditions on ice crystallisation in ice cream. *Journal of Food Engineering*, 39(2): 179–191.

Saberi, A.H., O.-M. Lai, and J.F. Toro-Vázquez. 2011. Crystallization kinetics of palm oil in blends with palm-based diacylglycerol. *Food Research International*, 44(1): 425–435.

Sadeghpour, A., M.L. Parada, J.L. Vieira, M. Povey, and M. Rappolt. 2018. Global small-angle X-ray scattering data analysis of triacylglycerols in the molten state (part I). *The Journal of Physical Chemistry B*, 122(45): 10320–10329.

Sánchez-García, Y.I., S.K. Bhangu, M. Ashokkumar, and N. Gutiérrez-Méndez. 2018a. Sonocrystallization of lactose from whey. *Technological Approaches for Novel Applications in Dairy Processing*, 51.

Sánchez-García, Y.I., K.S. García-Vega, M.Y. Leal-Ramos, I. Salmeron, and N. Gutiérrez-Méndez. 2018b. Ultrasound-assisted crystallization of lactose in the presence of whey proteins and κ-carrageenan. *Ultrasonics Sonochemistry*, 42: 714–722.

Sánchez-García, Y.I., N. Gutiérrez-Méndez, R.E. Orozco-Mena, V.H. Ramos-Sánchez, and M.Y. Leal-Ramos. 2019. Individual and combined effect of pH and whey proteins on lactose crystallization. *Food Research International*, 116: 455–461.

Sangwal, K. 2007. *Additives and Crystallization Processes: From Fundamentals to Applications*. John Wiley & Sons.

Santhalakshmy, S., S.J. Don Bosco, S. Francis, and M. Sabeena. 2015. Effect of inlet temperature on physicochemical properties of spray-dried jamun fruit juice powder. *Powder Technology*, 274: 37–43.

Sato, K. 2018. *Crystallization of Lipids: Fundamentals and Applications in Food, Cosmetics, and Pharmaceuticals*. John Wiley & Sons.

Saw, M.H., E. Hishamuddin, C.L. Chong, C.B. Yeoh, and W.H. Lim. 2017. Effect of polyglycerol esters additive on palm oil crystallization using focused beam reflectance measurement and differential scanning calorimetry. *Food Chemistry*, 214: 277–284.

Schoeman, L., P. Williams, A. du Plessis, and M. Manley. 2016. X-ray micro-computed tomography (μCT) for non-destructive characterisation of food microstructure. *Trends in Food Science & Technology*, 47: 10–24.

Silva, R., J. Lee, V. Gibon, and S. Martini. 2017. Effects of high intensity ultrasound frequency and high-speed agitation on fat crystallization. *Journal of the American Oil Chemists' Society*, 94(8): 1063–1076.

Singh, K., S.P. Gupta, A. Kumar, and A. Kumar. 2019. The effect of high intensity ultrasound (HIU) on the kinetics of crystallization of sucrose: Elimination of latent period. *Ultrasonics Sonochemistry*, 52: 19–24.

Sonwai, S., and D. Rousseau. 2010. Controlling fat bloom formation in chocolate – Impact of milk fat on microstructure and fat phase crystallisation. *Food Chemistry*, 119(1): 286–297.

Syamaladevi, R.M., K.N. Manahiloh, B. Muhunthan, and S.S. Sablani. 2012. Understanding the influence of state/phase transitions on ice recrystallization in Atlantic Salmon (Salmo salar) during frozen storage. *Food Biophysics*, 7(1): 57–71.

Thuy, N.T.H., A. Kongkaew, A. Flood, and A. Boontawan. 2017. Fermentation and crystallization of succinic acid from Actinobacillus succinogenes ATCC 55618 using fresh cassava root as the main substrate. *Bioresource Technology*, 233: 342–352.

Torbica, A., D. Jambrec, J. Tomić, B. Pajin, J. Petrović, S. Kravić, and I. Lončarević. 2016. solid fat content, pre-crystallization conditions, and sensory quality of chocolate with addition of cocoa butter analogues. *International Journal of Food Properties*, 19(5): 1029–1043.

Truong, T., M. Palmer, N. Bansal, and B. Bhandari. 2017. Effect of solubilised carbon dioxide at low partial pressure on crystallisation behaviour, microstructure and texture of anhydrous milk fat. *Food Research International*, 95: 82–90.

Ullah, J., P.S. Takhar, and S.S. Sablani. 2014. Effect of temperature fluctuations on ice-crystal growth in frozen potatoes during storage. *LWT - Food Science and Technology*, 59(2), Part 1: 1186–1190.

Van der Sman, R. 2017. Predicting the solubility of mixtures of sugars and their replacers using the Flory–Huggins theory. *Food &Function*, 8(1): 360–371.

van Westen, T., and R.D. Groot. 2018. Predicting the kinetics of ice recrystallization in aqueous sugar solutions. *Crystal Growth & Design*, 18(4): 2405–2416.

Verstringe, S., S. Danthine, C. Blecker, F. Depypere, and K. Dewettinck. 2013. Influence of monopalmitin on the isothermal crystallization mechanism of palm oil. *Food Research International*, 51(1): 344–353.

Verstringe, S., K. Moens, N. De Clercq, and K. Dewettinck. 2015. Crystallization behavior of monoacylglycerols in a hydrophobic and a hydrophilic solvent. *Food Research International*, 67: 25–34.

Wagh, A., M.K. Walsh, and S. Martini. 2013. Effect of lactose monolaurate and high intensity ultrasound on crystallization behavior of anhydrous milk fat. *Journal of the American Oil Chemists' Society*, 90(7): 977–987.

Wang, F., Y. Liu, Q. Jin, J. Huang, Z. Meng, and X. Wang. 2011. Kinetic analysis of isothermal crystallization in hydrogenated palm kernel stearin with emulsifier mixtures. *Food Research International*, 44(9): 3021–3025.

Wang, T., H. Lu, J. Wang, Y. Xiao, Y. Zhou, Y. Bao, and H. Hao. 2017. Recent progress of continuous crystallization. *Journal of Industrial and Engineering Chemistry*, 54: 14–29.

West, R., and D. Rousseau. 2019. Regression modelling of the impact of confectioner's sugar and temperature on palm oil crystallization and rheology. *Food Chemistry*, 274: 194–201.

Widlak, N., R.W. Hartel, and S. Narine. 2001. *Crystallization and Solidification Properties of Lipids*. The American Oil Chemists Society.

Wong, S.Y., R.K. Bund, R.K. Connelly, and R.W. Hartel. 2010. Modeling the crystallization kinetic rates of lactose via artificial neural network. *Crystal Growth & Design*, 10(6): 2620–2628.

Xanthakis, E., A. Le-Bail, and H. Ramaswamy. 2014. Development of an innovative microwave assisted food freezing process. *Innovative Food Science & Emerging Technologies*, 26: 176–181.

Xia, J., and D.A. Case. 2012a. Sucrose in aqueous solution revisited, Part 1: Molecular dynamics simulations and direct and indirect dipolar coupling analysis. *Biopolymers*, 97(5): 276–288.

Xia, J., and D.A. Case. 2012b. Sucrose in aqueous solution revisited, Part 2: Adaptively biased molecular dynamics simulations and computational analysis of NMR relaxation. *Biopolymers*, 97(5): 289–302.

Xiao, Y., Y. Liu, Y. Wang, Y. Jin, X. Guo, Y. Liu, X. Qi, H. Lei, and H. Xu. 2020. Heat-induced whey protein isolate gels improved by cellulose nanocrystals: Gelling properties and microstructure. *Carbohydrate Polymers*, 231: 115749.

Xu, Y., X. Zhao, Q. Wang, Z. Peng, and C. Dong. 2016a. Thermal profiles, crystallization behaviors and microstructure of diacylglycerol-enriched palm oil blends with diacylglycerol-enriched palm olein. *Food Chemistry*, 202: 364–372.

Xu, Y., C. Wei, X. Zhao, C. Lu, and C. Dong. 2016b. A comparative study on microstructure, texture, rheology, and crystallization kinetics of palm-based diacylglycerol oils and corresponding palm-based oils. *European Journal of Lipid Science and Technology*, 118(8): 1179–1192.

Yuan, Y., S. Hong, H. Lian, K. Zhang, and H. Liimatainen. 2020. Comparison of acidic deep eutectic solvents in production of chitin nanocrystals. *Carbohydrate Polymers*, 236: 116095.

Zakharov, B., A. Fisyuk, A. Fitch, Y. Watier, A. Kostyuchenko, D. Varshney, M. Sztucki, E. Boldyreva, and E. Shalaev. 2016. Ice recrystallization in a solution of a cryoprotector and its inhibition by a protein: Synchrotron X-ray diffraction study. *Journal of Pharmaceutical Sciences*, 105(7): 2129–2138.

Zamanipoor, M.H., and R.L. Mancera. 2014. The emerging application of ultrasound in lactose crystallisation. *Trends in Food Science & Technology*, 38(1): 47–59.

Zeng, F., F. Ma, Q. Gao, S. Yu, F. Kong, and S. Zhu. 2014. Debranching and temperature-cycled crystallization of waxy rice starch and their digestibility. *Carbohydrate Polymers*, 113: 91–96.

Zhang, G., W. Xie, and D. Wu. 2020a. Selective localization of starch nanocrystals in the biodegradable nanocomposites probed by crystallization temperatures. *Carbohydrate Polymers*, 227: 115341.

Zhang, H., Y. Chen, S. Wang, L. Ma, Y. Yu, H. Dai, and Y. Zhang. 2020b. Extraction and comparison of cellulose nanocrystals from lemon (Citrus limon) seeds using sulfuric acid hydrolysis and oxidation methods. *Carbohydrate Polymers*, 238: 116180.

Zhang, J., Y. Meng, J. Wu, J. Qin, W. Hui, T. Yao, and S. Yu. 2020c. Monitoring sugar crystallization with deep neural networks. *Journal of Food Engineering*. doi: 10.1016/j.jfoodeng.2020.109965:109965.

Zhang, Z., D.W. Sun, Z. Zhu, and L. Cheng. 2015. Enhancement of crystallization processes by power ultrasound: Current state-of-the-art and research advances. *Comprehensive Reviews in Food Science and Food Safety*, 14(4): 303–316.

Zhao, J.-H., Y. Ding, Y. Nie, H.-W. Xiao, Y. Zhang, Z. Zhu, and X.-M. Tang. 2016. Glass transition and state diagram for freeze-dried Lentinus edodes mushroom. *Thermochimica Acta*, 637: 82–89.

Zhao, J.-H., F. Liu, X. Wen, H.-W. Xiao, and Y.-Y. Ni. 2015. State diagram for freeze-dried mango: Freezing curve, glass transition line and maximal-freeze-concentration condition. *Journal of Food Engineering*, 157: 49–56.

Zhou, Y., and Y.H. Roos. 2012. Stability and plasticizing and crystallization effects of vitamins in amorphous sugar systems. *Journal of Agricultural and Food Chemistry*, 60(4): 1075–1083.

Zulkurnain, M., F. Maleky, and V. Balasubramaniam. 2016. High pressure processing effects on lipids thermophysical properties and crystallization kinetics. *Food Engineering Reviews*, 8(4): 393–413.

8 Application of Ohmic Heating in Food Processing

M. Kumar
Department of Food Technology, Guru Jambheshwar University Science & Technology, Hisar, Haryana, India

A. Barmanray
Department of Food Technology, Guru Jambheshwar University Science & Technology, Hisar, Haryana, India

K.K. Dash
Department of Food Processing Technology, Ghani Khan Choudhury Institute of Engineering and Technology, Malda, West Bengal, India

S. Chakraborty
Department of Food Engineering and Technology, Tezpur University, Tezpur, Assam, India

CONTENTS

8.1 INTRODUCTION

Heating is one of the oldest methods of processing and preservation of foods. Application of popular novel advanced technologies such as ohmic heating, dielectric heating (which includes microwave heating and radio frequency heating), and inductive heating has brought major changes in the heat-based processing of food. Due to heat generation inside the food, these advanced technologies are highly energy efficient.

Ohmic heating is one of the most efficient novel thermal technologies to be applied in food processing. It is also known as Joule heating, electrical resistance heating, direct electrical resistance heating, electro-heating or electro-conductive heating technology. Ohmic heating is accomplished by passing electric current through the food materials, where heat is generated due to the effect of electrical resistance (Alwis and Fryer, 1990). The principle of ohmic heating is different from microwave and inductive heating in several ways: Ohmic heating allows food materials to be kept between the electrodes; the waveform is unrestricted; and the frequency applied is less than the radio or microwave frequency range.

In the nineteenth century, a successful application of electricity in food processing, the Electro-pure process, was developed for the pasteurization of milk (Getchel, 1935). Due to high processing costs, this application was discouraged (Fryer and Li, 1993), and as the inert materials required for the electrodes were in short supply, other applications were abandoned. However, several researchers have reported recent applications of ohmic heating in vegetables and meat products, flours and starches, etc. (Palaniappan and Sastry, 1991; Wang and Sastry, 1997; Castro et al., 2003; An and King, 2007). With proper maintenance of color and nutritional value, ohmic heating helps in the processing of food products with extended shelf life.

8.2 PRINCIPLES OF OHMIC HEATING

The ohmic heating principle depends on the properties of electrically conducting materials containing various charged particles, such as electrons in metals and ions or molecules (proteins) in food material. When an electric potential is applied to these conductors, electric current starts to flow, and vibration occurs throughout the food material. This results in the increase of temperature due to generation of heat (Sastry et al., 2014). The motion of charged particles towards the electrodes of opposite polarity constitutes the electric current through the material. Both alternating (AC) and direct (DC) current can be used. The principle of ohmic heating is illustrated in Figure 8.1.

Ohmic heat generation can be characterized by the following equation (Sastry et al., 2014):

$$Q = E^2 \times \sigma \tag{8.1}$$

Here, Q is heat generation rate per unit volume (w/m^3), E refers to electric field strength (volt per unit length V/m) and σ stands for electrical conductivity (S/m).

Electrical conductivity can be expressed by the following equation (Zareifard et al., 2014):

$$\sigma = \left(\frac{1}{R}\right)\left(\frac{L}{A}\right) \tag{8.2}$$

Here, R is resistance (ohm), L is distance between electrodes (m) and A is area of the electrodes (m^2).

During ohmic heating, voltage gradient and electrical conductivity of the material affect the rate of energy generation. Successful application of ohmic heating in any food material can be achieved if the electrical conductivity is sufficient. Hence

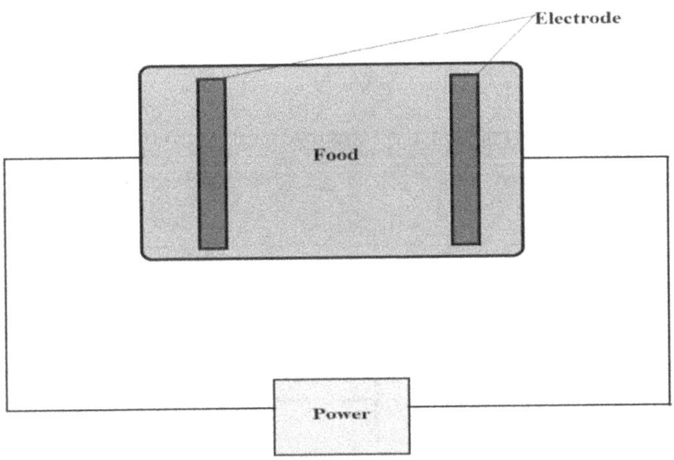

FIGURE 8.1 Ohmic heating working principle and operation

materials which do not conduct electricity, for example non-polar materials (oil, alcohol or sugar) cannot be heated by the application of ohmic heating. Therefore product electrical conductivity is the main governing factor for the applicability of ohmic heating. However, ohmic heating can be applied to most food materials as they contain a significant amount of water and charged ions, viz. acids, salts, or proteins. During treatment, electrical conductivity of the food materials decreases since drag on the ionic movement reduces with the decrease of viscosity of the materials that is achieved as they encounter the increased temperature (Ruan et al., 2002a; Knirsch et al., 2010).

At higher temperature, heating becomes more efficient because of the increase in conductivity attained with a constant voltage control. Material with lower conductivity becomes more conductive at higher temperature. As electrical conductivity is affected by the concentration of ionic constituents, it is possible to alter the conductivity by the infusion of salts. Regardless of the mode of heating used, the σ-T relationship is linear in the case of liquids. Furthermore, due to the effect of non-polar constituents in the pulp, electrical conductivity decreases with the increase of pulp content. In the case of fluids containing particulate material, the equivalent rate of heat generation can be achieved with similar electrical conductivity for the two phases (Ruan et al., 2002a).

For particulate food mixtures, HTST (high-temperature short-time) ohmic heating is thus an alternative to conventional HTST. At a specific temperature, resistance of the material and the geometry of the ohmic heating chamber affect electrical conductivity. The relationships between σ and T for solid and liquid materials are expressed by the following mathematical equations (Sastry et al., 2014; Ruan et al., 2002a). Figure 8.2 shows flow of current in particulates and liquid of a suspension.

$$\sigma_T = \sigma_o\left[1+mT\right]\left(\text{Liquid materials}\right) \tag{8.3}$$

$$\sigma_T = \sigma_r\left[1+m\left(T-T_r\right)\right]\left(\text{Solid materials}\right) \tag{8.4}$$

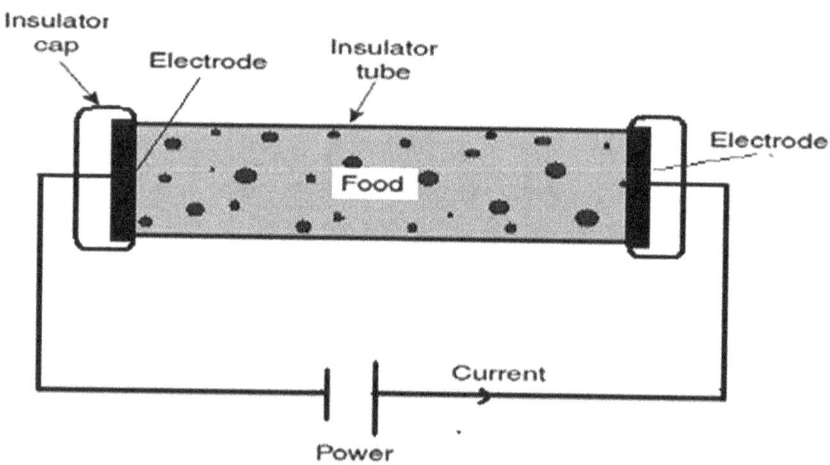

FIGURE 8.2 Flow of current in particulates and liquid of a suspension

Here σ_T refers to electrical conductivity at temperature T, σ_0 refers to electrical conductivity of material at 0°C, σ_r refers to electrical conductivity at any reference temperature T_f, m stands for temperature coefficient of electrical conductivity (°C^{-1}), and T is temperature (°C) at which σ is to be calculated.

8.3 FACTORS AFFECTING THE OHMIC HEATING PROCESS

The ohmic heating rate of a food material is impacted by different factors like electrical conductivity, field quality, molecule size, fixation, ionic focus, and anodes. Electrical conductivity is the most significant factor for the process modeling of the ohmic heating cycle (Fryer and Li, 1993). Food materials with more than one phase are influenced by these parameters on the basis of effective conductivity of the mixture. Moreover, heating rates of different constituents are influenced by particle orientation and geometry. The ohmic heating rate of a suspension of elongated particles is greater when the applied electric field is parallel to the shorter axis of the particles.

The factors associated with the ohmic heating process can be categorized into two main groups: process-dependent factors and food composition-dependent factors.

8.3.1 Process-Dependent Factors

Among the process-dependent factors, temperature, electric field strength, voltage gradient, location of the particles and electrode types are significant.

8.3.1.1 Electric Field Strength

Electric field intensity significantly influences the electrical conductivity of the food material. Increased electric field intensity affects electrical conductivity, and eventually a raised heating rate can be observed for the food material, and higher rates of microbial inactivation and extraction can be attained. For effective microbial inactivation, generally electric field strength in the range of 104–105 V/cm is proposed. With field intensity under 100 V/cm, extraction and expressing of various food materials can be accomplished (Kaur and Singh, 2016).

8.3.1.2 Location of Particles

Particle orientation and position also play a vital role in the ohmic heating of food materials. Location of mass particles with respect to the electrodes influences their ohmic heating behavior. For the parallel connection, faster heating of liquid phase is observed as compared to solid phase, whereas in case of series connection reverse situation is observed (Kaur and Singh, 2016).

8.3.1.3 Electrode Types

To ensure effective ohmic heating, it is important to choose appropriate materials and dimensions for the cathode (Samaranayake et al., 2005). Different rates of heating are observed for the electrodes of different materials, for example, stainless steel, titanium, aluminum, and shapes (dimensions). The type of electrodes and product containment cells significantly influence heat losses from the periphery of the food materials during ohmic heating (Kaur and Singh, 2016).

8.3.1.4 Temperature

Quantification of electrical conductivities with varying temperature is an important aspect of research. Linear or quadratic relationships for the mapping of temperature dependency of various food products during ohmic heating are reported by various researchers. For example, a linear rising trend of electrical conductivity with the increase in temperature has been observed during heating of a carrot–starch mixture (Kaur and Singh, 2016). Similarly, during the ohmic heating of grape juice, a linear trend was observed for the deactivation kinetics of polyphenoloxidase (Kaur and Singh, 2016). In a study of beetroot, electrical conductivity increased more sharply with the variation of temperature in ohmic heating than conventional heating (Kaur et al., 2016).

8.3.1.5 Frequency of the Electric Current

In the case of fruit juices, direct proportionality between frequency and electrical conductivity is observed. For the ohmic heating of turnip, an inverse relationship of frequency and heating rate was observed (Kaur et al., 2016). In general ohmic heating rates are faster at lower alternating current frequencies.

8.3.2 Food Composition-Dependent Factors

8.3.2.1 Electrical Conductivity

Electrical conductivity is one of the most important factors in the heat generation rate (HGR) during ohmic heating. There is a direct relationship between HGR and the square of electric field strength and electrical conductivity. Table 8.1 shows electrical conductivity ranges for different food products during ohmic heating.

8.3.2.2 Particle Size and Concentration of Material

The ohmic heating rate is also affected by particle size and concentration of the material. As particle size expands, the electrical conductivity diminishes, and an extreme decline in the heating rate can be observed. The rate of heating is also influenced by the concentration of dissolved sugars (brix value) and level of the suspended particles in a liquid phase (Kaur and Singh, 2016).

8.3.2.3 Ionic Concentration

During ohmic heating of biological tissue, auxiliary changes in the tissue, such as cell wall proto-pectin breakdown, ejection of non-conductive gas air pockets, relaxation, and watering down of the consistency, cause incremental increases in the electrical conductivity, leading to expansion of ionic mobility. The heating rate can thus be seen to be directly proportional to the ionic concentration. Consequently, for powerful ohmic heating of the material, uniform ionic concentration is required (Kaur and Singh, 2016).

TABLE 8.1
Electrical Conductivity Ranges for Different Food Products during Ohmic Heating (Icier, 2012)

Type of Food	Temperature Range (°C)	Voltage Gradients Applied (V/cm)	Concentration Range	Electrical Conductivity (S/m)
Apple juice	30–80	20–60	20–60% v/v soluble solids	0.30–1.00
Sourcherry juice	30–75	20–60	20–60% v/v soluble solids	0.70–1.60
Orange juice	25–80	42.42	0–21% solid	0.50–1.25
	30–60	20–60	0.2–0.6 mass fraction	0.15–1.15
Grape juice	20–80	20–40	Fresh	0.38–0.78
Strawberry pulp				
Pulp 1	20–80	32–70	Brix 14.50, 2.5% w/w starch	0.20–0.50
Pulp 2		40–80	Brix 26.50, 2.5% w/w starch	0.10–0.35
Strawberry-apple sauce	20–100	25–70	Not reported	0.11–0.34
Apricot puree	20–60	20–70	12.79%	0.61–1.25
Peach puree	25–60	20–70	9.90%	0.51–0.91
Carrot puree–Starch mixture	20–80	12.5	30–50% puree in starch solution (4% starch, 0.5% salt)	0.70–1.80
Tomato juice	25–80	42.42	0–16.7% solid	0.70–1.65
Tomato soup	20–120, under pressure of 30 psi	0–20.5	Dilution of commercial condensed tomato soup in distilled water (1:1 v/v)	1.50–4.80
Tomato Sauce	25–80	12.5	75.3% w/w water, 0.60% w/w xanthan gum, 22.6% w/w sugar, 1.5% w/w NaCl	1.25–3.20
Milk	10–70	15–45	3% fat	0.30–1.00
Ice-cream mix	4–80	10–60		
Standard mix			9.8% fat	0.15–1.00
Maras-type mix			3.3% fat	0.18–1.00
Whole liquid egg	4–60	20	Fresh	0.32–0.83
Reconstituted whey solution	20–80	20–40	8–24% w/v solute	0.98–3.90
Potato starch suspension	30–90	20	20% w/w starch, 1% salt	1.65–2.90
	30–60	10	33.3% starch, 0.4% salt	0.50–0.84

8.4 APPLICATION OF OHMIC HEATING

Throughout history, conventional methods have been used for the processing and preservation of foods. The most important criteria remain retention of food quality and nutritional value of the final product. Degradation of quality and nutritional properties of the final products are the most significant demerits of conventional methods. In recent years, novel thermal processing methods resulting in lower levels of quality deterioration have emerged. Ohmic heating has been shown to be superior to conventional and other novel thermal techniques in terms of quality, nutritional value, and rate of processing. The following sections describe various applications of ohmic heating.

8.4.1 Application to Food Products

8.4.1.1 Starches and Flour

Starches are one of the most important constituents in food products, used as stabilizing additives, moisture retainers, and thickeners. Severe degradation of starch granules and increase in the soluble solids occurs in pre-gelatinized starches prepared using drum dryer and extrusion processes. Starches treated with ohmic heating show the lowest enthalpy measured by differential scanning calorimetry (DSC), that is, higher pre-gelatinization occurs. Starches ohmically treated at 20 V/cm are reported to show the lowest enthalpy. Furthermore, decreased gelatinization onset temperature and faster swelling can be observed in ohmically heated rice flour (Kaur and Singh, 2016).

8.4.1.2 Meat Products

The effect of ohmic heating on the thermo-physical properties of meat items has been investigated by several researchers. Thermo-physical properties of different meat emulsions can be studied adequately at 20, 40, 60, and 80°C. Addition of lean to fat meat shows increasing impact on the overall conductivity. Muscle fiber in beef biceps femoris shows a significant impact on dielectric properties. Generally, ohmic heating is viable for the heating of meat emulsion and batters. In processed beef, electrical conductivity rises with the ohmic heating rate because of the increased mobility of sodium chloride and phosphate ions. A higher electrical conductivity and diminished ohmic heating time is seen in the case of whole meats with lower endogenous fat or processed meats with the least amount of included fat. Alignment of beef fibers according to the current flow can greatly improve electrical conductivity (Kaur and Singh, 2016).

8.4.1.3 Fruits and Vegetables

Nutritional quality is affected during conventional processing of fruit and vegetable products. Ohmic heating is an alternative technology for obtaining fruit and vegetable products of high quality. Significant retention of food quality can be achieved with the application of minimal ohmic heating, and drying time can also be minimized. Faster deactivation of enzymes and microorganisms occurs due to the interactions of the electric field. The voltage of the ohmic heating set-up is directly proportional to the solid content of the materials in the degradation of heat-sensitive compounds (anthocyanins, ascorbic acid, and vitamin C). Minimal degradation at low voltage level can be achieved compared with the results with conventional heating methods. At high voltage there is greater degradation and the greatest degradation occurs during the application of ohmic heating at low electric field frequency. Electrochemical reactions, believed to cause the degradation, are minimized with treatment above 100 Hz, when similar degradation rates can be observed for both ohmic and conventional heating processes (Kaur et al., 2016).

8.4.2 Application in Food Processing

Ohmic heating is a new technique for the cooking of food, in which cooking time and energy consumption can be reduced while quality and nutritional properties of the final product are retained (Farid et al., 2014).

Continuous sterilization has replaced conventional batch methods as it is efficient, convenient, and safe. For HTST treatment of solid–liquid mixtures, it is difficult to shift from a batch to a continuous process. Ohmic heating is a promising way of aseptically processing fluids containing large particles, which cannot be achieved using other heating techniques. Ohmic heating is therefore suitable for the sterilization of a wide range of food products (Ruan et al., 2002b; Makroo et al., 2020).

In food processing blanching is generally used to inactivate enzymes that cause quality degradation of fruits and vegetables. The major disadvantage of conventional blanching methods is the longer processing time required. Ohmic heating-based blanching processing time is shorter, with rapid generation of heat inside the food material. Furthermore, for multiple-phase food material, ohmic heating-based blanching can be applied efficiently by inducing similar electrical conductivity for all phases (Icier and Bozkurt, 2014).

Thawing of food stuffs generally requires more time than freezing. The application of ohmic heating shortens conventional thawing time of food materials, because the electrical resistance of frozen food rapidly generates heat, resulting in rapid and volumetric thawing of the food products. This reduces drip losses from food products during thawing (Seyhun et al., 2014).

The efficiency and rate of extraction of solutes, dye, oil, and juices from vegetable tissue can be improved by the application of ohmic heating. The rate of extraction can be controlled by varying process parameters such as time, temperature, electric field strength (V/cm), and frequency of electric current (Kumar et al., 2014).

8.5 IMPACT OF OHMIC HEATING ON THE QUALITY PROPERTIES OF THE FOOD MATERIAL

The biggest challenge for the thermal processing of food is the retention of quality and nutritional properties. Ohmic heating helps to improve overall food quality by reducing processing time (Ruan et al., 2002a). Various researchers have reported that superior or similar quality is obtained in ohmically processed foods than is achieved by conventional methods (Knirsch et al., 2010; Kaur and Singh, 2016; Saxena et al., 2016). Further research is required to improve understanding the effect of ohmic heating on the nutritive, organoleptic, and functional properties of various foods. Ohmic heating applied to preheating, cooking, blanching, processing, and extraction of valuable compounds in the food industry showed superior retention of nutritional and organoleptic properties (Icier and Ilicali, 2005).

8.5.1 Impact on Color

Color is an important indicator when evaluating the impact of different thermal processing methods, including ohmic heating, on the degradation of functional properties of food material. Significant changes in functional compounds of the food material based on explicit variation of color are reported by various researchers. One such example is ohmic heating treatment of acerola, in which changes in anthocyanins and carotenoid pigments can be indicated by color change. In the treatment of acerola, the color parameters L*, a*, b* decrease over time. Another such example is the use of color as an index for the representation of carotenoid degradation during ohmic heating treatment of orange juice (Kaur et al., 2016).

8.5.2 Impact on Rheological Properties

When juice is treated in an ohmic heating unit, changes in rheological properties are an important quality indicator. Rheological characteristics of quince nectar treated with ohmic heating with an electric field strength of 10–40 V/cm at 50 Hz were reported, based on various mathematical models, by Bozkurt and Icier (2009). The Herschel–Bulkley model provided the best fit for the prediction of rheological behavior. Shear thinning behavior was observed in quince nectar at pasteurization conditions. No adverse effects on the rheological properties of the ohmically heated juice were observed.

8.5.3 Impact on Viscosity and pH

Ohmic heating treatment has a significant impact on viscosity and pH of the food material. Singh et al. (2008) reported viscosity of various fruit juices in the temperature range of 25–70°C. As the juice temperature increased, viscosity decreased. The reason for the decrease in viscosity is the gradual decrease in cohesive force occurring at higher temperatures. On the other hand, Darvishi et al. (2013) reported the impact of voltage gradient on the pH of pomegranate juice: as the voltage gradient increased, the pH decreased.

8.6 IMPACT OF OHMIC HEATING ON THE INACTIVATION OF ENZYMES

Various quality-deteriorating enzymes such as lipoxygenase, polyphenol oxidase, peroxidase, etc., may be present in food, either inherently or produced by microorganisms growing on it (Saxena et al., 2017). In every plant tissue, and in several fungi and bacteria, pectin methyl esterase enzyme is inherently available. Inactivating these enzymes is therefore the prime concern in the processing of most food materials (Kumar et al., 2014). Inactivation of these enzymes occurs at a faster rate in ohmic heating than conventional heating due to the application of electric fields (Kaur and Singh, 2016). First-order kinetics is followed for the inactivation of enzymes in both conventional and ohmic heating methods.

8.7 IMPACT OF OHMIC HEATING ON THE INACTIVATION OF MICROORGANISMS

Ohmic heating can be effectively applied in the preservation of food. At lower frequencies (50–60 Hz), ohmic heating can be applied to the electroporation of the cell wall of microbes, ultimately helping to minimize D-value more successfully than conventional thermal methods (Knirsch et al., 2010).

All living cells, whether a prokaryote or eukaryote, contain cell membranes. These membranes are composed of lipids (fatty molecules) and proteins. Many prokaryotes have an additional layer outside the membrane known as the cell wall. At the low frequencies (50–60 Hz) and high field strengths (>100 V/cm) most commonly associated with ohmic heating, the naturally porous cell walls can allow the cell membrane to build up charges, forming disruptive pores. Electroporation occurs because the cell membrane has a specific dielectric strength, which can be exceeded by the electric field. The dielectric strength of a cell membrane is related to the amount of lipids (acting as insulators) present in the membrane itself. The pores formed can vary in size depending on the strength of the electric field, and can reseal after a short time. Excessive exposure causes cell death due to the leakage of intracellular components through the pores. Therefore, electroporation is highly damaging to a cell and enhances the lethal effects of thermal abuse already present due to ohmic heating. Figure 8.3 illustrates the electroporation process of a cell during ohmic heating.

By reducing microbial load and destroying pathogenic microorganisms, ohmic heating maintains the safety and shelf life of food products. Conventional methods of thermal processing, such as sterilization and pasteurization, can be replaced by ohmic heating for microbial destruction of food production without affecting consumer acceptability. Microbial inactivation in relation to ohmic heating is primarily thermal in nature, much like conventional heating. Electrical pre-treatment by ohmic heating can reduce the intensity of additional thermal applications for subsequent microbial inactivation. Microbial inactivation curves of ohmic heating are similar to conventional heating curves except for a difference in the slope, which is probably due to the presence of an electric field. No particular pathogens with a unique resistance to ohmic heating have been discovered; thermal death kinetic studies available for pathogens and spoilage bacteria can therefore be followed when designing experiments and systems that utilize ohmic heating. It inactivates pathogenic microorganisms such as *E coli, Salmonella typhimurium*; *Listeria monocytogenes* that are below the detection limit for orange juice, tomato juice, apple juice, and skimmed milk (Lee et al., 2012; Park and Kang, 2013; Park et al., 2017). In the case of milk, ohmic heating can be applied for the degradation of microorganisms, with retention of most nutrients.

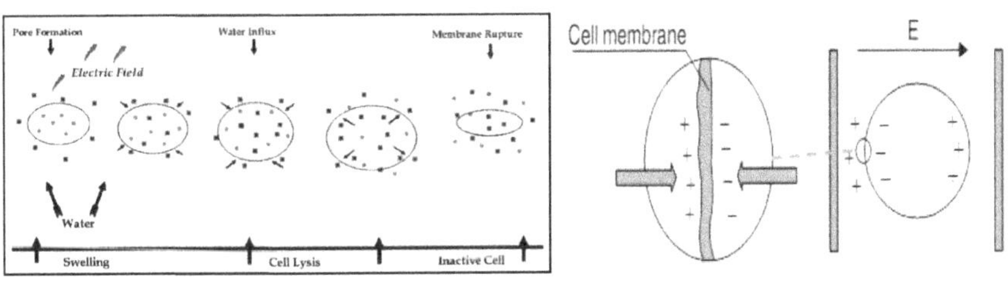

FIGURE 8.3 Electroporation process of a cell

8.8 RESPONSE SURFACE METHODOLOGY-BASED MATHEMATICAL MODELING OF THE OHMIC HEATING PROCESS

This section reports a case study on ohmic heating of strawberry. Strawberries are the most popular berries in the world (Mondal, 2010). Their nutritional composition varies by cultivar and variety, cultivation technique, area, climate, harvesting time, and ripeness (Hakala et al., 2003). Strawberries are used both fresh and in processed form, such as juice, jams, and jellies. It helps control blood sugar levels by slowing digestion. Strawberry is a rich source of nutrients, with distinctive color, flavor and taste (Giampieri et al., 2012).

Strawberry is available for a short time in April and May. Huge losses occur due to poor post-harvest handling. Blanching is a common practice used to inactivate surface micro-flora and some catalytic enzymes responsible for oxidation. Ohmic heating is a very rapid and uniform method of heating, so it is assumed that good-quality blanched products can be developed from ohmic blanching of strawberries. In the present work, field strength and temperature attained in ohmic blanching of strawberries (0.4% salt solution) were examined; vitamin C content, anthocyanin content, and heating rate were measured.

Electric field strength (x_1) and temperature (x_2) were the independent variables. The ranges and values of the experiment variables used in the coded and real form are given in Table 8.2. Central Composite Rotatable Design (CCRD) was used to optimize the independent variables (electric field strength and temperature) based on the dependent variables (heating rate, anthocyanin, and vitamin C). The experimental designs, which involved 13 experiments with five replications of the central point, are presented in Table 8.3 in coded and uncoded form. For the ohmic heating treatment, 100 g strawberry were immersed in 0.4% salt solution.

8.8.1 Analysis of Response for Strawberry

The present CCRD matrix models were used to interpret statistical data inferred from the model term values <0.0001 in the absence of fit values for heating rate, anthocyanin and vitamin C. The statistical data also showed reasonable agreement with the model, as the predicted R^2 values 0.88, 0.763 and 0.816 were nearly equal to the adjusted R^2 values 0.953, 0.922, and 0.964 for anthocyanin, vitamin C, and heating rate respectively.

TABLE 8.2
Values of Independent Variables at Five Levels of the Central Composite Rotatable Design (CCRD) for Ohmic Heating of Strawberry

Independent Variables	Code	Levels in Coded Form				
		−1.41	−1	0	+1	+1.41
Electric field strength	x_1	6.79	17.5	26.3	35.1	38.75
Temperature	x_2	55.86	60	70	80	84.14

TABLE 8.3
Central Composite Rotatable Design (CCRD) in Coded and Uncoded Levels of Variables Employed for Ohmic Heating of Strawberry

	Coded Variables		Uncoded Variables	
Experiment No.	x1	x2	x1	x2
1	−1.00	−1.00	17.5	60
2	1.00	−1.00	35.1	60
3	−1.00	1.00	17.5	80
4	1.00	1.00	35.1	80
5	−1.41	0.00	6.79	70
6	1.41	0.00	38.75	70
7	0.00	−1.41	26.30	55.86
8	0.00	1.41	26.3	84.14
9	0.00	0.00	26.3	70
10	0.00	0.00	26.3	70
11	0.00	0.00	26.3	70
12	0.00	0.00	26.3	70
13	0.00	0.00	26.3	70

The heating rate data given in Table 8.4 indicate non-significant values for linear term (temperature, 0.836), interactive term (0.548) and quadratic term for temperature (0.112), whereas only two terms, linear term (<0.0001) and quadratic term (0.004) for temperature show significant values. So simplification of the terms would improve modeling of the heating value response.

TABLE 8.4
Analysis of Variation in Heating Rate

Source	Coefficient of Model Terms	Sum of Squares	df	Mean Square	F Value	Prob>F
Constant	4.25					
x_1	2.36***	47.31	1	47.31	301.03	0.000
x_2	0.03^{ns}	0.01	1	0.01	0.05	0.836
$x_1.x_2$	0.125^{ns}	0.06	1	0.06	0.40	0.548
x_1^2	0.40***	2.77	1	2.77	17.60	0.004
x_2^2	0.28^{ns}	0.52	1	0.52	3.29	0.112
Complete model						
Regression		51.05	5	10.21	64.96	0.000
Lack of Fit		1.07	3	0.36	44.51	0.002
Pure Error		0.03	4	0.01		
Residual		1.10	7	0.16		
Total		52.15	12			
R^2		0.979		Adeq. Precision		27.528
Adjusted R^2		0.964				

ns: non-significant, df: degree of freedom
Level of significance: *P<0.1, **P<0.05, ***P<0.01

The p values of the Prob>F were significant for the linear component, and quadratic components for all variables except the electric field strength (X_1^2) for anthocyanin response (Table 8.5). The interactive components for the anthocyanin also show a non-significant value (0.081), clearly indicating higher levels of both variables will not enhance the response. This is also supported by the non-significant value of the quadratic component for electric field strength (0.055), indicating a reduction in the response value at higher temperature.

The model shows (Table 8.6) the significant value for linear terms 0.001 and < 0.001 for electric field strength and temperature respectively, interactive terms (0.002) and quadratic term for electric field strength (0.018), whereas the reported value for the quadratic coefficient of temperature was not significant (0.329) for vitamin C. The non-significant value of the temperature indicated a negative impact on the response (vitamin C). The electric field strength has shown a positive effect on the vitamin C with increasing level of the variables.

The models for the measured variables were as follows:

$$\text{Heating Rate} = 4.25 + 2.36{*}X_1 + 0.03{*}X_2 + 0.12{*}X_{12} + 0.40{*}X_1^2 + 0.28{*}X_2^2$$

$$\text{Anthocyanin} = 21.79 + 3.33{*}X_1 - 2.97{*}X_2 - 0.97{*}X_{12} - 0.53{*}X_1^2 - 1.14{*}X_2^2$$

$$\text{Vitamin C} = 19.84 + 2.38{*}X_1 - 3.06{*}X_2 - 2.80{*}X_{12} - 0.90{*}X_1^2 - 0.49{*}X_2^2$$

Figures 8.4 and 8.5 illustrate the impacts of electric field intensity and temperature on the heating rate of strawberry. The electric field strength tends to have a significant positive effect on the heating rate, while temperature does not. Comparable outcomes are reported in Table 8.4. Figures 8.6 and 8.7 show the effects of electric field

TABLE 8.5
Analysis of Variation for Anthocyanin

Source	Coefficient of Model Terms	Sum of Squares	df	Mean Square	F Value	Prob>F
Constant	21.79					
x_1	3.33***	93.80	1	93.80	102.32	0.000
x_2	−2.97***	70.39	1	70.39	76.79	0.000
$x_1.x_2$	−0.98*	3.80	1	3.80	4.15	0.081
x_1^2	−0.53*	4.83	1	4.83	5.27	0.055
x_2^2	−1.14**	8.82	1	8.82	9.62	0.017
Complete model						
Regression		229.41	5	45.88	50.05	0.000
Lack of Fit		2.41	3	0.80	0.80	0.554
Pure Error		4.01	4	1.00		
Residual		6.42	7	0.92		
Total		235.83	12			
R^2		0.973		Adeq. Precision		23.960
Adjusted R^2		0.953				

ns: non-significant, df: degree of freedom
Level of significance: *P<0.1, **P<0.05, ***P<0.01

TABLE 8.6
Analysis of Variance for Vitamin C Content

Source	Coefficient of Model Terms	Sum of Squares	df	Mean Square	F Value	Prob>F
Constant	19.84					
x_1	2.38***	47.87	1	47.87	32.46	0.001
x_2	−3.06***	75.12	1	75.12	50.94	0.000
$x_1.x_2$	−2.80***	31.36	1	31.36	21.27	0.002
x_1^2	−0.90**	13.98	1	13.98	9.48	0.018
x_2^2	−0.49ns	1.62	1	1.62	1.10	0.329
Complete model						
Regression		216.55	5	43.31	29.37	0.000
Lack of Fit		7.52	3	2.51	3.58	0.125
Pure Error		2.80	4	0.70		
Residual		10.32	7	1.47		
Total		226.87	12			
R^2		0.955		Adeq. Precision		20.070
Adjusted R^2		0.922				

ns: non-significant, df: degree of freedom
Level of significance: *P<0.1, **P<0.05, ***P<0.01

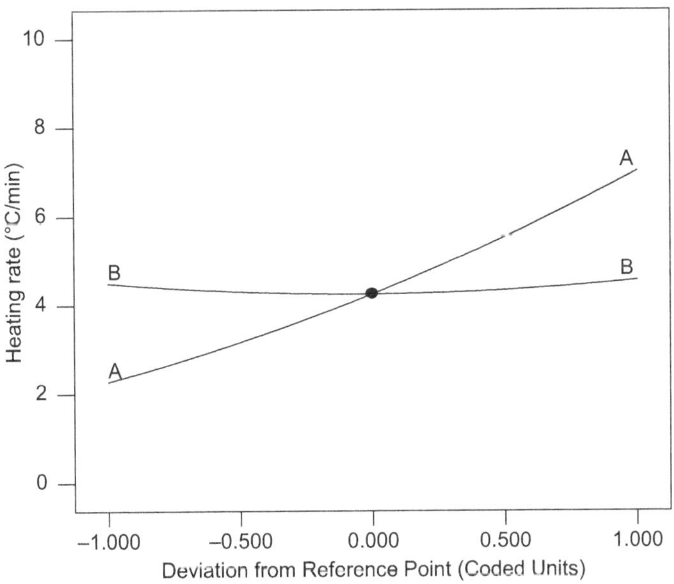

FIGURE 8.4 Perturbation plot of variables for the heating rate in strawberry

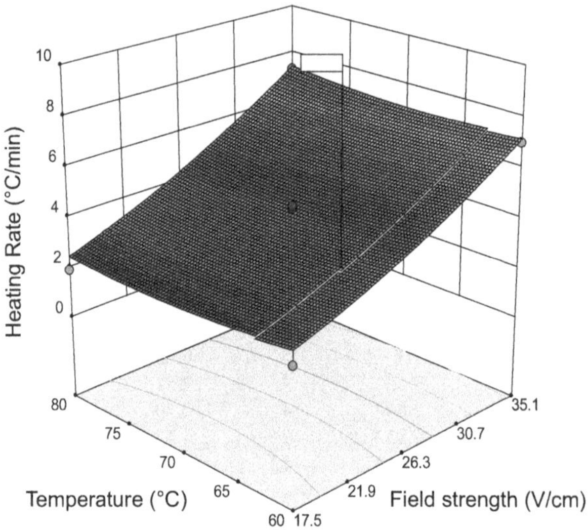

FIGURE 8.5 Effect of electric field strength and temperature attained on heating rate of whole strawberry

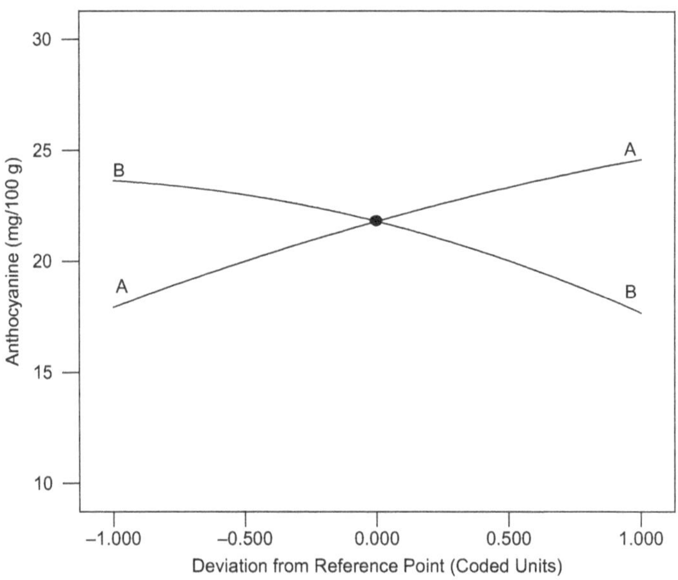

FIGURE 8.6 Perturbation plot of variables for the anthocyanin in strawberry

intensity and temperature on the anthocyanin content of strawberry. From these figures, the positive impact of electric field strength and negative impact of temperature on the anthocyanine content can be observed. The results reported in Table 8.5 show a similar impact. The impact of these factors on the vitamin C content of strawberry is illustrated in Figures 8.8 and 8.9. As with the anthocyanine content, the positive

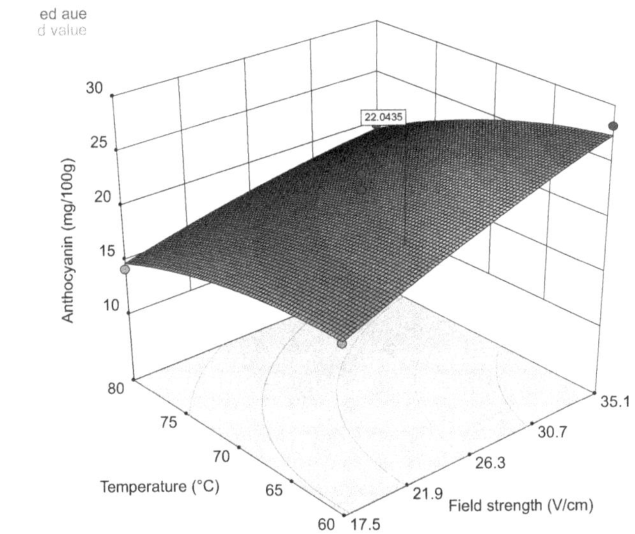

FIGURE 8.7 Effect of electric field strength and temperature attained on anthocyanin content of whole strawberry

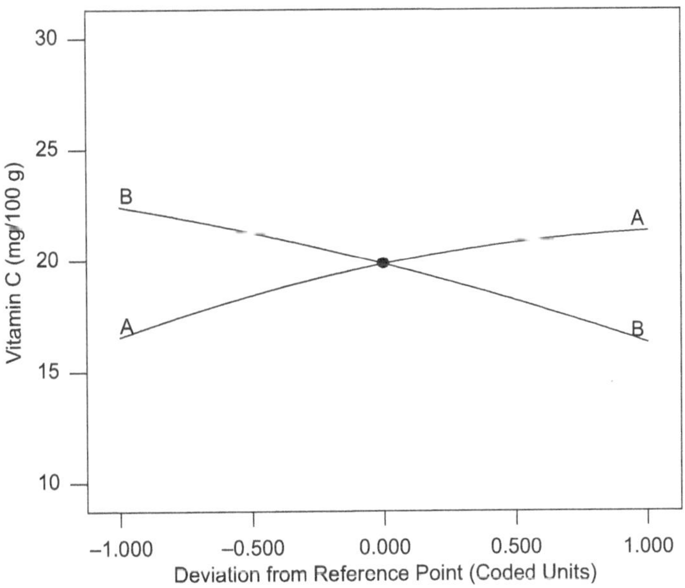

FIGURE 8.8 Perturbation plots of variables for the vitamin C content in strawberry

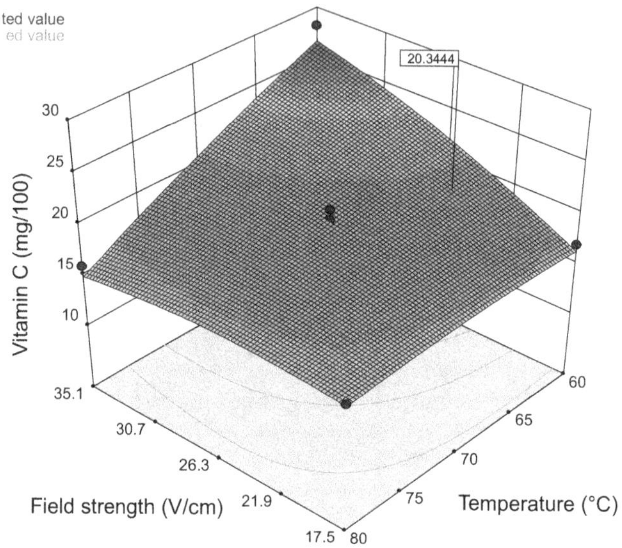

FIGURE 8.9 Effect of electric field strength and temperature attained on vitamin C content of whole strawberry

effect of electric field intensity and the negative effect of temperature on the vitamin C content of strawberry can be clearly seen. Similar impacts of these factors are reported in Table 8.6.

8.8.2 Optimization of Ohmic Heating Conditions

A numerical multi-response optimization technique with desirability function was used to estimate the optimum level of electric field strength and temperature. These optimum conditions (within the range constraint) were used to process the strawberry by ohmic heating and the predicted and actual values at the optimized condition were compared. The least deviation in actual values (heating rate: 7.13°C/min, anthocyanin: 26.5 mg/100 g and vitamin C content: 27.3 mg/100 g) against predicted values (heating rate: 7.10°C/min, anthocyanin: 26.3 mg/100 g and vitamin C content: 27.3 mg/100 g) was found at an electric field strength of 35.1 V/cm and temperature of 60°C. This was the best of three optimum conditions having less than 5% deviation.

8.9 MERITS AND DEMERITS OF OHMIC HEATING PROCESS

Ohmic heating frameworks are beneficial because investment is optimized, with expanded productivity, instant system shut-down, and reduced maintenance costs due to the absence of moving parts. In conventional heating, the time taken to raise the temperature at the cold point may over-process the rest of the particles and the encompassing fluid. This over-processing causes pulverization of supplements and lessening of flavor. Ohmic heating processes the particles and encompassing fluid at the same time, preventing overcooking.

The cost of a commercial ohmic heating framework, including establishment, can exceed US$9,000,000 USD. A food that has fat globules can be troublesome to successfully warm ohmically, as the fat globules are non-conductive. Even where these globules are within an exceptionally electrical conductive environment where current can flow around them, they may warm more slowly because of absence of electrical conductivity. Any pathogenic microscopic organisms in these globules may get less heat treatment than the remainder of the substance.

Another slight impediment is associated with the rise in electrical conductivity of a substance as its temperature rises. This presents the possibility of 'runaway' warming. An ohmic heating framework that has not been cleaned thoroughly enough may suffer electrical arcing because of protein deposition on the terminals. However, by using the information on the previously mentioned issues in planning an ohmic heating framework, these weaknesses can be controlled.

8.10 CONCLUSION

Ohmic heating (OH) is defined as a process wherein electric current is passed through materials with the primary purpose of heating them. It is one of the newest methods of heating foods. A large number of actual and potential applications exist for OH, including blanching, evaporation, dehydration, fermentation, extraction, sterilization and pasteurization processes. OH is an advanced thermal processing method wherein the food material, which serves as an electrical resistor, is heated by passing electricity through it. Electrical energy is dissipated into heat, which results in rapid and uniform heating.

This heating technique is also called electrical resistance heating, Joule heating, or electro-heating, and may be used for a variety of applications in the food industry. Foods that contain water and ionic salts are capable of conducting electricity but they also have a resistance which generates heat when an electric current is passed through them. The electrical resistance of a food is the most important factor in determining how quickly it will heat. Conductivity measurements are therefore made in product formulation, process control and quality assurance for all foods that are heated electrically. Electrical resistance of a food is measured using a multimeter connected to a conductivity cell.

REFERENCES

Achir, N., C. Dhuique-Mayer, T. Hadjal, K. Madani, J.P. Pain, and M. Manuel Dornier. 2016. Pasteurization of citrus juices with ohmic heating to preserve thecarotenoid profile. *Innov. Food Sci. Emer. Tech.* 33: 397–404.

Alwis, A.A.P., and P.J. Fryer. 1990. The use of network theory in the finite-element analysis of coupled thermo-electric fields. *Comm. Appl. Num. Meth.* 6: 61–66.

An, H.J., and J.M. King. 2007. Thermal characteristics of ohmically heated rice starch and rice flours. *J. Food Sci.* 72(1): C84–C88.

Assiry, A.M., S.K. Sastry, and C.P. Samaranayake. 2006. Influence of temperature, electrical conductivity, power and pH on ascorbic acid degradation kinetics during ohmic heating using stainless steel electrodes. *Bioelectrochem.* 68: 7–13.

Bozkurt, H., and F. Icier. 2009. Rheological characteristics of quince nectar during ohmic heating. *Int. J. Food Prop.*, 12, 844–859.

Castro, I., J.A. Teixeira, S. Salengke, S.K. Sastry, and A.A. Vicente. 2003. The influence of field strength, sugar and solid content on electrical conductivity of strawberry products. *J. Food Pro. Eng.* 26: 17–29.

Cho, W.I., J.Y. Yi, and M.S. Chung. 2016. Pasteurization of fermented red pepper paste by ohmic heating. *Innov. Food Sci. Emer. Tech.* 34: 180–186.

Darvishi, H., M.H. Khostaghaza, and G. Najafi. 2013. Ohmic heating of pomegranate juice: Electrical conductivity and pH change. *J. Saudi Soc. Agric. Sci.*, 12: 101–108.

Farid, M.M., W.J. Kong, and N. Ozkan. 2014. Ohmic cooking of food. In: H.S. Ramaswamy, M. Marcotte, S. Sastry, and K. Abdelrahim (eds.), *Ohmic Heating in Food Processing*. Boca Raton, FL: CRC Press. ISBN: 13:978-1-4200-7109-2 (eBook - PDF).

Fryer, P.J. and Z. Li. 1993. Electrical resistance heating of foods. *J. Food Sci. Tech.* 4: 364–369.

Getchel, B.E. 1935. Electric pasteurization of milk. *Agri. Eng.* 16(10): 408–410.

Giampieri, F., S. Tulipani, J.M.A. Suarez, J.L. Quiles, B. Mezzetti, and M. Battino. 2012. The strawberry: Composition, nutritional quality, and impact on human health. *Nutri.* 28: 9–19.

Icier, F. 2012. Ohmic heating of fluid foods. In: *Novel Thermal and Non-Thermal Technologies for Fluid Foods*. London: Academic Press, pp. 305–367).

Icier, F. and H. Bozkurt. 2014. Ohmic blanching. In: H.S. Ramaswamy, M. Marcotte, S. Sastry, and K. Abdelrahim (eds.), *Ohmic Heating in Food Processing*. Boca Raton, FL: CRC Press. ISBN: 13:978-1-4200-7109-2 (eBook - PDF).

Icier, F. and C. Ilicali. 2005. Temperature dependent electrical conductivities of fruit purees during ohmic heating. *Food Res. Inter.* 38(10): 1135–1142.

Kaur, N. and A.K. Singh. 2016. Ohmic heating: Concept and applications – A review. *Cri. Rev. Food Sci. Nutri.* 56(14): 2338–2351.

Kaur, R., Gul, K., and Singh, A.K. 2016. Nutritional impact of ohmic heating on fruits and vegetables – A review. *Cog. Food Agri.* 2(1): 1159000.

Knirsch, M.C., C.A. Dos Santos, A.A.M. de Oliveira Soares, and T.C.V. Penna. 2010. Ohmic heating–a review. *Tren. Food Sci. Tech.* 21(9): 436–441.

Kumar, J.P., M. Amanathan, and T.V. Ranganatha. 2014. Ohmic heating technology in food processing – A review. *Inter. J. Eng. Res. Tech.* 3(2): 1236–1241.

Kumari, P. and B.S. Khatkar. 2015. Physico-chemical properties and nutritional composition of aonla (*Emblicaofficinalis*) varieties. *Inter. Food Res. J.* 22(6): 2358–2363.

Lee, S., H. Sagong, S. Ryu, and D. Kang. 2012. Effect of continuous ohmic heating to inactivate Escherichia coli O157:H7, *Salmonella Typhimurium* and *Listeria monocytogenes* in orange juice and tomato juice. *J. App. Micro.* 723–731.

Lima, M. and S.K. Sastry. 1999. The effects of ohmic heating frequency on hot-air drying rate and juice yield. *J. Food Eng.* 41: 115–119.

Makroo, H.A., N.K. Rastogi, and B. Srivastava. 2020. Ohmic heating assisted inactivation of enzymes and microorganisms in foods: A review. *Tren. Food Sci. Tech.*

Ozkan, N., I. Ho, and M. Farid. 2004. Combined ohmic and plate heating of hamburger patties: Quality of cooked patties. *J. Food Eng.* 63: 141–145.

Palaniappan, S. and S.K. Sastry. 1991. Electrical conductivities of selected solid foods during ohmic heating. *J. Food Pro. Eng.* 14: 136–221.

Park, I., J. Ha, and D. Kang. 2017. Investigation of optimum ohmic heating conditions for inactivation of Escherichia coli O157:H7, Salmonella entericaserovarTyphimurium, and Listeria monocytogenes in apple juice. *BMC Micro.* 17(117): 1–8.

Park, I. and D. Kang. 2013. Effect of electropermeabilization by ohmic heating for inactivation of Escherichia coli O157:H7, Salmonella entericaserovartyphimurium, and Listeria monocytogenes in buffered peptone water and apple juice. *App. Env. Micro.* 79(23): 7122–7129.

Patil, S.S., A. Singh, M. Kulshrestha, B.K. Kumbhar, and M. Sahgal. 2011. Ohmic heating characteristics of aonla pulp. *Pantnagar J. Res.* 9(1): 120–126.

Ruan, R., et al. 2002a. Ohmic heating. In: P. Richardson (ed.), *Thermal Technologies in Food Processing*. Woodhead Publishing Ltd. ISBN: 1 85573 558.

Ruan, R., et al. 2002b. Ohmic heating. In: K.J. Henry and C. Chapman (eds.), *The Nutrition Hand Book for Food Processors*. Woodhead Publishing Ltd. ISBN: 1 85573 665 9 (e book).

Samaranayake, C.P., S.K. Sastri, and H. Zhang. 2005. Pulsed ohmic heating – A novel technique for minimization of electrochemical reactions during processing. *J. Food Sci.* 70(8): E460–E465.

Sastry, S.K., B.F. Heskitt, S.S. Sarang, R. Somavat, and K. Ayotte. 2014. Why ohmic heating? Advantages, applications, technology, and limitations. In: *Ohmic Heating in Food Processing* (Vol. 7). Boca Raton, FL: CRC Press.

Saxena, J., H. Ahmad Makroo, and B. Srivastava. 2017. Effect of ohmic heating on polyphenol oxidase (PPO) inactivation and color change in sugarcane juice. *J. Food Pro. Eng.* 40(3): e12485.

Saxena, J., H.A. Makroo, and B. Srivastava. 2016. Optimization of time-electric field combination for PPO inactivation in sugarcane juice by ohmic heating and its shelf life assessment. *LWT - Food Sci. Tech.* 71: 329–338.

Sengun, I.Y., P. Kendirci, and F. Icier. 2013. Effect of ohmic treatment on quality characteristic ofmeat: A review. *Meat Sci.* 93(3): 441–448.

Seyhun, N., S.G. Sumnu, and H.S. Ramaswamy. 2014. Ohmic heating as thawing and tempering technology. In: H.S. Ramaswamy, M. Marcotte, S. Sastry, and K. Abdelrahim (eds.), *Ohmic Heating in Food Processing*. Boca Raton, FL: CRC Press. ISBN: 13:978-1-4200-7109-2 (eBook - PDF).

Singh, H., S.P. Singh, and T.P. Singh 2008. Study on viscosity and electrical conductivity of fruit juices. *J. Food Sci. Technol.* 45: 371–372.

Wang, W.C. and S.K. Sastry. 1997. Starch gelatinization in ohmic heating. *J. Food Eng.* 34: 225–242.

Yokozawa, T., H.Y. Kim, H.J. Kim, T. Okubo, D.C. Chu, and L.R. Juneja. 2007. Amla (*Emblicaofficinalis*Gaertn.) prevents dyslipidaemia and oxidative stress in the ageing processtheayurvedic wonder. *J. Bas. Clin. Physio. Pharma.* 21(1): 93–105.

Zareifard, M.R., M. Marcotte, H.S. Ramaswamy, and Y. Karimi. 2014. Electrical conductivity: Importance and methods of measurement. In: H.S. Ramaswamy, M. Marcotte, S. Sastry, and K. Abdelrahim (eds.), *Ohmic Heating in Food Processing*. Boca Raton, FL: CRC Press. ISBN: 13:978-1-4200-7109-2 (eBook - PDF).

9 Extraction Technology Applications in Food and Biological Products

Qurat Ul Eain Hyder Rizvi
Department of Food Technology, Eternal University, Baru Sahib, Himachal Pradesh, India

Rafeeya Shams
Division Food Science and Technology, Sher-e-Kashmir University of Agricultural Sciences & Technology, Jammu, India

Aamir Hussain Dar
Department of Food Technology, Islamic University of Science and Technology, Awantipora, Kashmir, India

Ishrat Majid
Department of Food Technology, Islamic University of Science and Technology, Awantipora, Kashmir, India

Anurag Singh
Department of Food Science and Technology, National Institute of Food Technology and Entrepreneurship Management, Kundli, Haryana, India

Shafat Khan
Department of Food Technology, Islamic University of Science and Technology, Awantipora, Kashmir, India

Mohsin Bashir Aga
Department of Food Technology, Islamic University of Science and Technology, Awantipora, Kashmir, India

CONTENTS

9.1 INTRODUCTION

Functional foods are designed from raw materials using various processes such as fortification, enrichment, or other alterations to nutritional components (Cukelj et al., 2016). Foods having functional properties are the equivalent of conventional foods in all respects except that they contain the added constituent of various active biological compounds (Granato et al., 2017). Essential for the authenticity and traceability of engineered foods is appropriate composition, food structure, and stability (Granato et al., 2018). The raw materials involved are minimally processed fresh fruits (Putnik et al., 2016), vegetables (Montesano et al., 2018), and plants that add flavor or have therapeutic benefits (Kovačević et al., 2018), together with their waste and supplementary products (Putnik et al., 2017e, 2017f, 2017g).

Current consumer trends in the "wellness" sector indicate that food products from plant constituents are considered healthy, fresh, and possessing great nutritional benefits, but consumers also demand sustainable processing and production (Putnik et al., 2017a, 2017b, 2017c, 2017d). Today's functional foods incorporate plant extracts rich in biologically active compounds (BACs), extracted using traditional methods that frequently have a negative thermal impact on quality and yield (Huang et al., 2013) as well as being time consuming, requiring large quantities of organic solvents and significant energy expenditure. However, there are alternative, sustainable extraction methods, including microwave-assisted extraction (MAE), ultrasound-assisted extraction (UAE), high-pressure assisted extraction (HPAE), high voltage electric discharges assisted extraction (HVED), pulsed electric fields assisted extraction (PEF), supercritical fluids extraction (SFE) (Gabric et al., 2018; Koubaa et al., 2017). These extraction processes are associated with green concepts of manufacturing for optimum consumption while also keeping energy consumption to a minimum (Herrero et al., 2010; Chemat et al., 2017).

The main aim of the extraction process is to separate essential components from by-products using specific solvents. Crude extracts acquired using this technique contain a composite blend of various plant metabolites: phenolics, glycosides, alkaloids, flavonoids, and terpenoids. Fluid extracts and tinctures are used as medicinal agents. Antioxidants can be extracted from by-products of the food industry. Vegetables and fruits are essential sources of carotenoids, ascorbic acid, and phenolic compounds. Various bioactive components that have tremendous health benefits can be extracted from natural sources. The interaction of these components with DNA, proteins, and other bioactive components can be exploited to prepare remedial compounds derived from natural food products (Ajikumar et al., 2008). These novel biological compounds (NBCs) are key components of drugs or health supplements, although due to limited natural sources, they are mostly developed in relatively modest amounts as mixtures or conjugates, and the extraction process is complex and labor-intensive (Lam, 2007). The complexity and diverse structural framework of the molecules concerned makes for less profitable chemical synthesis (Lam, 2007). New bioengineering techniques and processes to enhance detection, production, and segregation of molecular species have expanded the range of NBCs, providing a chance to extract compounds of essential use (Wang and Weller, 2006). Such techniques offer a novel approach to enhance the yield of highly purified products that are beneficial in a number of applications. Examples include molecular biological methods to utilize microbes (bacteria) to form drugs such as isoprenoid, components obtained from vegetation (Chang and Keasling, 2006), and novel dihydroflavonols and flavanones (Harvey, 2008).

Polyphenols are basically organic molecules with large molecular-weight mostly found in plant sources. They have various health benefits, due to their antioxidant properties, and are mostly used in prevention of cardiovascular, neurodegenerative and inflammatory diseases, and cancer. The latest research focuses on enhancing the extraction of polyphenols using various techniques such as microwave, ultrasound, pulsed electric field, pressurized liquid, high hydrostatic pressure, and supercritical fluid. This chapter discusses the benefits and limitations of techniques to assist extraction and compares them with conventional extraction processes and purification techniques with appropriate examples.

9.2 EXTRACTION METHODS

The health benefits of various components (in extract or isolated form) as functional food or as nutraceuticals make low-cost extraction and segregation of polyphenols essential. Various techniques are used to enhance the isolation and production of NBCs. While several studies have been devoted to characterization of phenolic components and biological activities, their extraction methods are mostly overlooked. Comparative benefits and drawbacks of various extraction processes are shown in Table 9.1.

9.2.1 Solvent Extraction

Solvent extraction is a conventional method of extracting many natural bioactive components, especially phenolic components, from a variety of sources (Gil-Chávez et al. 2013). Plant materials pretreated, for example by washing and drying, are subjected to various solvents like water, ether, hexane, chloroform, benzene, acetonitrile, methanol, and ethanol, which take up the essential molecules. The choice of solvent should exclude those which bind to proteins or carbohydrates (Robbins, 2003), and depends on the phenolic compound's location within the plant. The efficacy of this process depends on polarity of the solvent, solvent type, phenolic nature (degree of polymerization), and the interaction of other components (Naczk and Shahidi, 2004). The extraction of polyphenolic components was initially done using the solid–liquid extraction method, usually referred to as lixiviation or leaching (Luque-de-Castro and García-Ayuso, 1998). The material is placed in an extraction thimble placed in a container between a distillation flask and a condenser. Solvent distills and is condensed onto the extraction thimble container. When the solvent in the container reaches a certain level, it is returned by siphon action into the distillation flask. The whole process repeat still extraction is complete.

TABLE 9.1
Benefits and Drawbacks of Various Extraction Processes

Process	Advantages	Disadvantages	References
Solvent extraction	Easy to handle and rapid	Thermal degradation; high solvent consumption; long treatment time	Gil-Chávez et al. (2013)
MAE	Low solvent consumption; high extraction yield; cost-effective compared with solvent extraction; shorter extraction time than UAE	Expensive apparatus and equipment; complex operation compared with UAE; less eco-friendly because of organic solvents. Unsuitable for heat-labileh biomolecules; microwave energy absorbed by extraction solvent. Filtration is necessary.	Zhang et al. (2011)
PEF	Quick non-thermal process; shorter extraction time; improved yield. Facilitation of purified extract (grinding, reducing). Less effect on environment than traditional methods; low energy consumption.	Not well known. Essential to maintain high-energy inputs, process parameters, temperature treatment. Mechanism and intensification process is difficult.	Barba et al. (2015)
PSE		Possible degradation of thermo labileanalytes; greater equipment cost. 10 g max sample analysis limit.	Dobias et al. (2010)
SCF	Low solvent consumption; no filtration necessary; minimal wastage because recycling and reuse of supercritical fluid is possible.	Many parameters to optimize; costly. Unsuitable for pharmaceutical and drug specimens; unable to dissolve polar molecules	Abbas et al. (2008)
UAE	Easy to use; low power and energy requirement; greater product yield; less chemical usage; shorter processing time.	Higher solvent consumption; filtration essential; nominal power of the device. Proper optimization of ultrasound frequency, input power, engendering of cycle, and framework geometry required for exceptional yield. Structural geometry essential to achieve excellent results.	Barba et al. (2015)

This technique is used to obtain various components from materials such as polymers, bacteria, algae, fungi, microalgae, sediments, and plants (Plaza et al., 2010a). A preprocessed raw constituent is subjected to various solvents, in which the desired components (colorings and flavors) are dissolved (Starmans and Nijhuis, 1996). Centrifugation and filtration of samples eliminates residual matter. Polar and non-polar solvents like phenolics, alkaloids, aryl hydrocarbons, organochlorine pesticides, oils, and fatty acids are used to extract several components (Villa-Rodríguez et al., 2011). A large quantity of solvent requiring concentration for recovery needs to be employed. Some solvents require careful handling due to their toxicity for the environment and persons working with them, and extraction is time consuming (Miron et al., 2011). If the extract is for food purpose it is essential that the solvent should be completely removed from the mixture (Starmans and Nijhuis, 1996). The process of segregation by solvent extraction has shown improved results due to the latest developments and other processes like Soxhlet, microwave, ultrasound, and SFE (Szentmihályi et al., 2002).

9.2.2 Pressurized Liquid Extraction

NBCs can also be extracted from natural sources using pressurized liquid extraction (PLE) (Miron et al., 2011). PLE requires solvents from 1450 to 2175 psi and 50 to 200°C to enhance the extraction rate (Dunford et al., 2010). The latest methods offer rapid extraction obtaining higher yields and using a lower amount of solvents than the traditional solvent extraction method. When a GRAS (Generally Recognised as Safe)-status solvent or water is used for extraction, the extracted material becomes food grade (Plaza et al., 2010b).

PLE efficiently extracts various compounds such as flavonoids, saponins, and essential oils from TCM (Traditional Chinese Medicine) (Xu et al., 2015). Temperature-susceptible components can be extracted by this method as it requires minimum time. During the recovery of antioxidants from grape, a Maillard reaction was observed at 200°C (Vergara-Salinas et al., 2013). Gizir et al. (2008) used PLE to extract anthocyanin from black carrots.

Pressurized solvents are highly flexible because characteristics including density, dielectric constant, viscosity, and diffusivity can be regulated by altering the extraction system temperature and pressure (Pronyk and Mazza, 2009). Extraction by pressurized liquid enables rapid extraction utilizing less solvent; it has been used efficiently for anthocyanin extraction from several plant sources (Santos et al., 2012). Machado et al. (2015) used PLE with water, ethanol, and acidic water having pH 2.5 to extract antioxidants from cranberry waste at various temperatures (60, 80, and 100°C), comparing outcomes with traditional removal techniques (Soxhlet and maceration). The desired outcomes were achieved by PLE at 100°C, with ethanol and water. This technique has also been found to recover bioactive components efficiently from pomace. Xu et al. (2016) used PLE to examine the antiradical action of polysaccharides obtained from gooseberry. The best conditions were found to be 52°C and 1.6 MPa for 51 min. Machado et al. (2015) used PLE for extraction of antioxidants from blackberry pomace.

9.2.3 Supercritical Fluid Extraction

A supercritical fluid is a solvent heated above a certain temperature, known as the critical temperature, at which point it no longer has the properties of a liquid but is denser than a gas (Wang and Weller, 2006). Supercritical fluid extraction (SFE) is environmentally friendly, due to its use of natural solvents to extract compounds from plants, algae, microalgae, pomace, and from agricultural waste (Herrero et al., 2006). Its rapid speed and high selectivity are further advantages (Wang and Weller, 2006). SFE depends on various characteristics of the fluids, such as dielectric constant, diffusivity, viscosity, and density. Supercritical fluid has superior mobility to fluids as it relies on its density which, unlike liquid solvents, can be adjusted by altering temperature and pressure (Herrero et al., 2006).

SFE is widely utilized in the food industry. A number of products on the market are described as supercritical carbon dioxide processed products, including crude palm oil or rice bran oil, fatty acids, essential oils, wheat germ oil, green coffee oil, other bioactive compounds and phospholipids from vegetables and fruits. SFE is also used to purify and segregate compounds. SC-CO_2 (co-solvent) with water can be employed to extract caffeine only, without disturbing antioxidants. Supercritical extraction is also widely used in the food industry to add value to by-products. Wu et al. (2009) extracted polyphenols from rice wine lees and Abbasi et al. (2008) from pomegranate seeds, Shi et al. (2009) extracted carotenoids from tomato, and Cossuta et al. (2007) extracted carotenoids from sea buckthorn pomaces.

9.2.4 Microwave-Assisted Extraction

Microwaves are electromagnetic radiation of wavelength 1 mm to 1 m. Microwave energy can be absorbed into a medium such as water or foodstuffs where it becomes heat (Zhang et al., 2011). This principle can be used in microwave-assisted extraction (MAE). Water present in the cells of foodstuffs is heated and evaporated, generating a large pressure on the cell wall. MAE can be used to recover many NBCs, in particular those having antioxidant capacity such as carotenoids (Pasquet et al., 2011), phenolic compounds (Simsek et al., 2012), saponins, alkaloids, and terpenoids (Zhang et al., 2011). Most research concludes that MAE reduces extraction time and solvent consumption, improves antioxidant capacity, and provides higher or equivalent yields than other techniques. High antioxidant activity and high yields compared to rotary extraction were obtained in citrus mandarin peel extracts (Hayat et al., 2009), from the skin of peanuts (Ballard et al., 2010), from tomatoes (Li et al., 2011), and from onions (Zill-e-Huma et al., 2011). Microwaves are also a non-contact heat source which provides high-efficiency heating, enhances energy transfer, and reduces thermal gradient. Numerous classes of components, including antioxidants, flavorings, essential oils, and other organic components and pigments, can be segregated effectively using this technique (Li et al., 2013). Simsek et al. (2012) found that in both cherry pomace and *Raix Astragali* (herb), the amount of phenolic component yielded using MAE was the same as with traditional methods (rotatory and Soxhlet method) in the shortest time. When Beejmohun et al. (2007) obtained phenolic compound linseed by traditional extraction and by MAE, the latter produced a larger yield. Leadbeater (2014) reported that microwave equipment can be successfully used, as at higher temperatures it is simple, safe, and can be used several times, yield increases, reaction time reduces, and product purity is enhanced in comparison with traditional processes that require heating and the addition of solvent (Oroian and Escriche, 2015). Li et al. (2013) report extraction of bioactive components by microwaves without the use of solvents. They concluded this alternative method had advantages in terms of solvent use, energy utilization, CO_2 emissions, and reduced time.

9.2.5 Pulsed Electric Field-Assisted Extraction

Pulsed electric field-assisted extraction (PEF) relies on electroporation in the cell membrane. Time is reduced and yield is enhanced because mass transfer is increased during extraction due to disturbance of all the membrane structures. An electric field is carried through the cell membrane and the molecules inside it separate according to their charge. Pores are formed and permeability is increased through repulsion (Azmir et al., 2013; Rajha et al., 2015). PEF is also used to extract intracellular compounds from plant and agricultural waste. From a sustainability and economic perspective, PEF is a valuable technique for obtaining vital components present in various vegetable and fruit tissues, chiefly because of its capacity to disrupt and soften cell membranes, promoting the release of substances present within cells (Rosello-Soto et al., 2015).

PEF has the potential to extract lipophilic components during organic synthesis, as its mass transfer rate increases during extraction due to cell membrane disruption, resulting in reduced extraction time and increased yield (Toepfl et al., 2006). With reduced extraction time, there is an increase in the yield because PEF enhances discharge of intracellular components from plant tissue due to electroporation. A moderate electric field strength (500–1000 V/cm for 10^{-4} to 10^{-2} s) damages plant cell membranes with only a slight increase in temperature (Fincan and Dejmek, 2002). It is therefore regarded as a non-thermal method that can be used for temperature-labile components (Ade et al., 2001).

9.2.6 High Hydrostatic Pressure

High hydrostatic pressure (HHP) is a novel, green, non-thermal technique used as an alternative to thermal processes, principally to avoid microbial contamination and unwanted alteration in the product characteristics (Escobedo-Avellaneda et al., 2011). Pressure in the range 100–1000 MPa is used for treatment (Briones-Labarca et al., 2015). It is recommended by the USFDA, since it only requires electric power and does not generate waste (Andrés et al., 2016). To preserve compounds and enhance the shelf life of a large number of foods, associated pasteurization takes place. Currently, an enormous number of studies focus on novel uses of HHP, such as increased extraction of polyphenols. Jun (2009) compared HHP with traditional methods of polyphenol extraction from green tea. They used various solvents, liquid:solid ratios, and pressure parameters for their analysis comparing ultrasound, traditional, and HHP methods, concluding that the polyphenol yields could be raised by 15–30% by an increase in pressure from 100–600 MPa. Briones-Labarca et al. (2015) showed that HHP achieved more efficient extraction of bioactive components from *Vasconcelle apubescens* than other methods. They also found that atmospheric pressure resulted in less infusion of the bioactive components than HHP. Andrés et al. (2016) segregated bean protein at 70–90°C and at varying pressures (200, 400, and 600 MPa), using HHP. The functional characteristics were enhanced compared with less energetic methods (70°C and 200 MPa). HHP increases the number of phenolic components extracted from oak chips. The antioxidant activity and phenolic components of wine rise after treatment with oak chips (Tao et al., 2016). Andrés et al. (2016) analyzed chemical and physical characteristics, and also the bioactive components from HPP-treated smoothies made from soy. They found that the antioxidant content of the smoothies was significantly enhanced.

9.2.7 Ultrasound-Assisted Extraction

The ultrasound spectrum runs from 20 kHz to 100 MHz. Ultrasound creates cavitation through the production, growth, and collapse of bubbles (Azmir et al., 2013). Ultrasound-assisted extraction (UAE) can produce numerous analytes from various types of sample. Mass and heat transfer are accelerated by the effect of ultrasound through disruption in the cell walls, which leads to increased discharge of components (Rosello-Soto et al., 2015). UAE relies on two physical phenomena; rinsing the content of cells after the walls break down; and diffusion through the cell wall. The action of ultrasound can be regulated by frequency, temperature, sonication time, and pressure (Rajha et al., 2015; Vinatoru, 2001). UAE is an efficient technique for extracting bioactive components from industrial waste. Herrera and Luque-de-Castro (2004) used a 0.8 s duty cycle for 30 s to extract polyphenols like naringenin, ellagic acid, rutin, kaempferol, naringin, and quercetin from strawberries using a semi-automatic technique that relies on ultrasound. Phenolic and anthocyanin components were extracted from grape peel (Ghafoor et al., 2009), and process optimization was performed to find the best extraction conditions (temperature, solvent, and time). Research by Cho et al. (2006) recovered resveratrol from grapes using UAE. The yield obtained, despite reduced extraction time, was up by 24% and 28% in Gerbong and Campbell grapes respectively. El-Darra et al. (2013) also reported a considerable increment (up by 7%) in tannins and anthocyanins, and change in color and antioxidant activity in Cabernet Franc grapes after processing at 24 kHz for 5–15 min in 121–363 kJ/kg conditions.

9.2.8 Enzyme-Assisted Extraction

Enzyme-assisted extraction (EAE) can assist in breaking down the structure of cell walls, membranes, and micelles consisting of biomolecules like proteins and long-chain carbohydrates, whilst minimizing denaturation and clotting of proteins at elevated temperatures. EAE is effective due to the hydrolytic activity of enzymes on the cell wall, membrane, and inside the cell biomolecules, which assists the discharge of natural products. Cellulase, pectinase, and α-amylase are used in EAE. Most antioxidants are present in plant tissues. The cell walls of plants consist of polysaccharides, including pectin, cellulose, and hemicellulose, which prevent release of intracellular components. β-gluconase, xylanase, pectinase, and β-glucosidase interact with these polysaccharides and degrade the cell wall structure, allowing more intracellular components to be released (Moore et al., 2006; Chen et al., 2010). These enzymes are therefore used to optimize the extraction of components from plants (Wang et al., 2010). In the case of algae, rigid structure and thick cell walls are an obstruction, decreasing extraction efficiency of intracellular bioactive components (Wang et al., 2010).

Chandini et al. (2011) found that the highest polyphenol content was obtained when tannase was employed separately, and that the enzymes pectinase and tannase independently increase the quality of black tea extract. β-glucosidase, pectinase, and cellulose were used to extract luteolin and apigenin from pigeon pea leaves. In another experiment, flavonoids were extracted from *Ginkgo biloba*, which resulted in decimation of the cell wall and enhanced component solubility in the extract (Chen et al., 2010). EAE has also been used to increase the antioxidant yield from black carrot juice, and to extract vegetable oils (Khandare et al., 2010). EAE was also used to extract gallic acid from agricultural waste for food additives (Curiel et al., 2010). Propyl gallate and pyrogallol are used in the manufacture of the antibacterial drug Trimethoprim (Curiel et al., 2010).

9.2.9 Steam Distillation and Hydro Distillation

Steam and hydro distillation (SD and HD) are used to extract volatile oil. Various natural components disintegrate in SD and HD. The antibacterial activity and chemical composition of primary and secondary essential oils from bergamot-mint (*Mentha citrata Ehrh.*) were considerably affected by the distillation method. The primary and secondary essential oils yielded using HDs were greater than those using conventional methods (Verma et al., 2016). Yahya and Yunus (2013) found that patchouli oil quality is influenced by extraction time. The substance of many compounds is either boosted or brought down when the extraction time increases.

9.3 APPLICATIONS OF EXTRACTION PROCESSES

- In ultrasound-assisted extraction, cost is reduced because extraction time is shorter, power usage is efficient, more concentrated extract is obtained at comparable S/L (solid–liquid) ratio, and the yield is better. Conventional maceration cannot be used at commercial level for effective antioxidant extraction as it is too energy- and time-consuming. The conditions selected for the ultrasound extraction pilot study were at the ideal state of ultrasound energy density (23 W/cm^2) and at temperature 36°C. It was concluded that a time of 30 min is required for successful extraction of material, comparable to that obtained by the traditional maceration process; this extraction time is then standardized for continuous production of antioxidants. Solvent material yield from the ultrasound extraction pilot system was similar to that from the cluster method for the same length of extraction time (21.7% return).
- PEF technology has been widely used in the food-processing industry both in Europe and the USA since the 1990s (Toepfl et al., 2006). Currently food products treated with PEF are produced commercially in Germany, the Netherlands,

and the UK in systems with 2000–8000 L/h capacity (Mchugh and Toepfl, 2016). A juice firm in Germany has a 10 t/h PEF industrial system which is used to increase quantity of apple juice (Muller et al., 2007). In potato, carrot, and sugar beet, changes in texture after PEF treatment were observed (Janositz and Knorr, 2010). The technique is currently used in many industrial installations to replace the customary pre-warming of potatoes (60°C, 30 min) in the manufacture of French fries (Mchugh and Toepfl, 2016).

- Microwave extraction without solvent is an eco-friendly technique for the removal of essential oils from fragrant herbs that are widely used in the food business (Filly et al., 2014). This dry distillation technique is performed without any water or added solvent. For the extraction of essential oil from rosemary (*Rosmarinus officinalis* L), solvent-free microwave extraction and HD (Clevenger apparatus) are comparable methods. MAE was used to segregate volatile from non-volatile components in Peumus boldus leaves (Petigny et al., 2014). Extract concentration was enhanced in comparison with the conventional method (equal liquid/solid ratio), and extraction time was reduced.
- Using low-polarity high-pressure water has numerous advantages over traditional extraction techniques. Successful applications include essential oils extraction from numerous plants (Mortazavi et al., 2010), biophenols extracted from *Oleaeuropaea* leaves (Japon-Lujana and Luque-de-Castro, 2006), lactones extracted from kava roots, sweet components extracted from *Siraitia grosvernorii*, and antioxidant components extracted from *Spirulina platensis* and *Ginkgo biloba* (Ibáñez et al., 1999; Ibáñez et al., 2003). The oil quality is better than that obtained through steam distillation, as steam distillation causes greater oxidation and gives a lesser content of terpenes. Even though not all terpenes are extracted, the yield is also somewhat greater than that obtained from steam refining.
- SC-CO_2 liquid extraction is currently used on a large scale for the extraction of valuable items from natural substances. Careful selection of supercritical solvent pressure and temperature can give an optimum yield of bioactive health-promoting compounds with the highest level of antiradical activity (Yi et al., 2009).

9.4 CONCLUSION

With increasing consumer awareness of the compounds present in food products, those obtained from natural sources are preferred to those obtained by chemical synthesis, largely because the latter are believed to have adverse effects. Consumer demand worldwide for natural food components is based not only on their functional properties but also on their positive impact on the environment and long-term human health compared to synthetic food components. Biotechnological extraction tools used to produce NBCs are already having a positive impact and show promise for the future. Numerous optimized extraction techniques to extract natural components from animal and plant sources have been developed. Research seeking to improve extraction yields while reducing processing time and the environmental damage caused by noxious solvents confirms that the substitution of green techniques for traditional methods is promising. The use of several technologies, both conventional and advanced, in combination, is likely to produce a higher yield in an eco-friendly and less time-consuming way.

It is clear that the future of NBCs is in the field of nutraceuticals and functional foods, which are gaining importance due to concerns over health and changes in eating habits. BAC (Biologically Active Compound) sources are found in abundance in vegetables, fruits, and therapeutic plants, and in food-processing waste. This waste is a sustainable raw material capable of achieving the synthesis of nutritious functional foods both economically and in an environmentally friendly way.

REFERENCES

Abbas, K.A., A. Mohamed, A.S. Abdulamir, and H.A. Abbas. 2008. A review on supercritical fluid extraction as new analytical method. *American Journal of Biochemistry and Biotechnology*, 4: 345–353.

Abbasi, H., K. Rezaei, Z.E. Mamdjomeh, and S.M.E. Mousavi. 2008. Effect of various extraction conditions on the phenolic contents of pomegranate seed oil. *European Journal of Lipid Science and Technology*, 110: 435–440.

Ajikumar, P., K. Tyo, S. Carlsen, O. Mucha, T. Phon, and G. Stephanopoulos. 2008. Terpenoids: Opportunities for biosynthesis of natural product drugs using engineered microorganisms. *Molecular Pharmaceutics*, 5(2): 167–190.

Andrés, V., L. Mateo-Vivaracho, E. Guillamón, M.J. Villanueva, and M.D. Tenorio. 2016. High hydrostatic pressure treatment and storage of soy-smoothies: Colour, bioactive compounds and antioxidant capacity. *LWT - Food Science and Technology*, 69: 123–130.

Azmir, J., I.S.M. Zaidul, M.M. Rahman, K.M. Sharif, A. Mohamed, F. Sahena, M.H.A. Jahurul, K. Ghafoor, N.A.N. Norulaini, and A.K.M. Omar. 2013. Techniques for extraction of bioactive compounds from plant materials: A review. *Journal of Food Engineering*, 117: 426–436.

Ballard, T.S., P. Mallikarjunan, K. Zhou, and S. O'Keefe. 2010. Microwave-assisted extraction of phenolic antioxidant compounds from peanut skins. *Food Chemistry*, 120(4): 1185–1192.

Barba, F.J., E. Puertolas, M. Brncic, I.N. Panchev, D.A. Dimitrov, and V. Athes-Dutour. 2015. Emerging extraction. In: C. Galanakis (ed.), *Food Waste Recovery: Processing Technologies and Industrial Techniques*. London: Academic Press, pp. 249–272.

Barba, F.J.P. Putnik, D. BursaćKovačević, M.M. Poojary, S. Roohinejad, J.M. Lorenzo, and M. Koubaa. 2017. Impact of conventional and non-conventional processing on prickly pear (Opuntia spp.) and their derived products: From preservation of beverages to valorization of by-products. *Trends in Food Science and Technology*, 67: 260–270.

Beejmohun, V., O. Fliniaux, E. Grand, F. Lamblin, L. Bensaddek, P. Christen, J. Kovensky, M.A. Fliniaux, and F. Mesnard. 2007. Microwave-assisted extraction of the main phenolic compounds in flaxseed. *Phytochemical Analysis*, 18(4): 275–282.

Briones-Labarca, V., M. Plaza-Morales, C. Giovagnoli-Vicuna, and F. Jamett. 2015. High hydrostatic pressure and ultrasound extractions of antioxidant compounds, sulforaphane and fatty acids from Chilean papaya (*Vasconcelleapubescens*) seeds: Effects of extraction conditions and methods. *LWT - Food Science and Technology*, 60(1): 525–534.

Chandini, S.K., L.J. Rao, M. Gowthaman, D. Haware, and R. Subramanian. 2011. Enzymatic treatment to improve the quality of black tea extracts. *Food Chemistry*, 127(3): 1039–1045.

Chang, M. and J. Keasling. 2006. Production of isoprenoid pharmaceuticals by engineered microbes. *Nature Chemical Biology*, 2(12): 674–681.

Chemat, F., N. Rombaut, A. Meullemiestre, M. Turk, S. Périno, A.S. Fabiano-Tixier, and M. Abert-Vian. 2017. Review of green food processing techniques: Preservation, transformation, and extraction. *Innovative Food Science and Emerging Technologies*, 41: 357–377.

Chen, S., X.H. Xing, J.J. Huang, and M.S. Xu. 2010. Enzyme-assisted extraction of flavonoids from Ginkgo biloba leaves: Improvement effect of flavonoltransglycosylation catalyzed by Penicillium decumbens cellulase. *Enzyme and Microbial Technology*, 48(1): 100–105.

Chen, Y., M.Y. Xie, and X.F. Gong. 2007. Microwave-assisted extraction used for the isolation of total triterpenoidsaponins from Ganodermaatrum. *Journal of Food Engineering*, 81: 162–170.

Cho, Y.J., J.Y. Hong, H.S. Chun, S.K. Lee, and H.Y. Min. 2006. Ultrasonication-assisted extraction of resveratrol from grapes. *Journal of Food Engineering*, 77: 725–730.

Choi, S.K., J.H. Kim, Y.S. Park, Y.J. Kim, and H.I. Chang. 2007. An efficient method for the extraction of astaxanthin from the red yeast *Xanthophyllomycesdendrorhous*. *Journal of Microbiol Biotechnology*, 17(5): 847–852.

Cossuta, D., B. Simandi, J. Hohmann, F. Doleschall, and T. Keve. 2007. Supercritical carbon dioxide extraction of sea buckthorn (*Hippophaerhamnoides L.*) pomace. *Journal of the Science of Food and Agriculture*, 87: 2472–2481.

Cravotto, G., L. Boffa, S. Mantegna, P. Perego, M. Avogadro, and P. Cintas. 2008. Improved extraction of vegetable oils under high-intensity ultrasound and/or microwaves. *Ultrasonics Sonochemistry*, 15: 898–902.

Cukelj, N., P. Putnik, D. Novotni, S. Ajredini, B. Voucko, and C. Duska. 2016. Market potential of lignans and omega-3 functional cookies. *British Food Journal*, 118: 2420–2433.

Curiel, J.A., L. Betancor, B. de las Rivas, R. Munoz, J.M. Guisan, G. Fernandez-Lorente. 2010. Hydrolysis of tannic acid catalyzed by immobilized–stabilized derivatives of tannase from *Lactobacillus plantarum*. *Journal of Agriculture and Food Chemistry*, 58(10): 6403–6409.

Dobias, P., P. Pavlıkova, M. Adam, A. Eisner, B. Benova, and K. Ventura. 2010. Comparison of pressurised fluid and ultrasonic extraction methods for analysis of plant antioxidants and their antioxidant capacity. *Open Chemistry*, 8: 87–95.

Dunford, N., S. Irmak, and R. Jonnala. 2010. Pressurised solvent extraction of policosanol from wheat straw, germ and bran. *Food Chemistry* , 119(3): 1246–1249.

El-Darra, N., N. Grimi-Eugene, V. Nicolas, and L.R. Maroun. 2013. Extraction of polyphenols from red grape pomace assisted by pulsed ohmic heating. *Food and Bioprocess Technology*, 6: 1281–1289.

Escobedo-Avellaneda, Z., M.P. Moure, N. Chotyakul, J.A. Torres, J. Welti-Chanes, and C.P. Lamela. 2011. Beneficios y limitacionesdelprocesamiento de alimentosportecnologías de altapresión: Efectosencomponentesfuncionales y contaminantesabioticos (Benefits and limitations of food processing by high-pressure technologies: Effects on functional compounds and abiotic contaminants). *CyTA-Journal of Food*, 9(4): 351–364.

Filly, A., X. Fernandez, M. Minuti, F. Visinoni, G. Cravotto, and F. Chemat. 2014. Solvent free microwave extraction of essential oil from aromatic herbs from laboratory to pilot and industrial scale. *Food Chemistry*, 150: 193–198.

Fincan, M. and P. Dejmek. 2002. In situ visualization of the effect of a pulsed electric field on plant tissue. *Journal of Food Engineering*, 55: 223–230.

Gabric, D., F. Barba, S. Roohinejad, S.M.T. Gharibzahedi, M. Radojcin, P. Putnik, and D. BursaćKovačević. 2018. Pulsed electric fields as an alternative to thermal processing for preservation of nutritive and physicochemical properties of beverages: A review. *Journal of Food Process Engineering*, 41(1): e12638.

Ghafoor, K., Y.H. Choi, J.Y. Jeon, and I.H. Jo. 2009. Optimization of ultrasound-assisted extraction of phenolic compounds, antioxidants and anthocyanins from grape (Vitisvinifera) seeds. *Journal of Agricultural and Food Chemistry*, 57: 4988–4994.

Ghafoor, K., T. Hui, and Y.H. Choi. 2011. Optimization of ultrasound-assisted extraction of total anthocyanins from grape peel. *Journal of Food Biochemistry*, 35: 735–746.

Gil-Chávez, G.J., A. Jose, J. Villa, J. Fernando Ayala-Zavala, B. Heredia, D. Sepulveda, M. Elhadi Yahia, and A. Gustavo González-Aguilar. 2013. Technologies for extraction and production of bioactive compounds to be used as nutraceuticals and food ingredients. *Comprehensive Reviews in Food Science and Food Safety*, 12(1): 5–23.

Gizir, A.M., N. Turker, and E. Artuvan. 2008. Pressurized acidified water extraction of black carrot [Daucuscarotassp. sativus var. atrorubensAlef.] anthocyanins. *European Food Research and Technology*, 226(3): 363–370.

Granato, D., D.S. Nunes, and F.J. Barba. 2017. An integrated strategy between food chemistry, biology, nutrition, pharmacology, and statistics in the development of functional foods: A proposal. *Trends in Food Science and Technol*, 62: 13–22.

Granato, D., P. Putnik, D.B. Kovacevic, J.S. Santos, V. Calado, R.S. Rocha, A.G.D. Cruz, B. Jarvis, O.Y. Rodionova, and A. Pomerantsev. 2018. Trends in chemometrics, food authentication, microbiology, and effects of processing. *Comprehensive Reviews in Food Science and Food Safety*, 17(3): 663–677.

Handa, S.S., S.P.S. Khanuja, G. Longo, and D.D. Rakesh. 2008. *Extraction Technologies for Medicinal and Aromatic Plants*. United Nations Industrial Development Organization and the International Centre for Science and High Technology.

Harvey, A. 2008. Natural products in drug discovery. *Drug Discovery Today*, 13(19–20): 894–901.

Hattab, M., G. Culioli, L. Piovetti, S. Chitour, and R. Valls. 2007. Comparison of various extraction methods for identification and determination of volatile metabolites from the brown alga Dictyopterismembranacea. *Journal of Chromatography*, 1143(1–2): 1–7.

Hayat, K., S. Hussain, S. Abbas, U. Farooq, B.M. Ding, S.Q. Xia, C.S. Jia, X.M. Zhang, and W.S. Xia. 2009. Optimized microwave-assisted extraction of phenolic acids from citrus mandarin peels and evaluation of antioxidant activity in vitro. *Separation and Purification Technology*, 70(1): 63–70.

Herrera, M.C. and M.D. Luque-de-Castro. 2004. Ultrasound-assisted extraction for the analysis of phenolic compounds in strawberries. *Analytical and Bioanalytical Chemistry*, 379: 1106–1112.

Herrero, M., A. Cifuentes, and E. Ibaez. 2006. Sub and supercritical fluid extraction of functional ingredients from different natural sources: Plants, food-by-products, algae and microalgae: A review. *Food Chemistry*, 98(1): 136–148.

Herrero, M., M. Plaza, A. Cifuentes, and E. Ibáñez. 2010. Green processes for the extraction of bioactives from Rosemary: Chemical and functional characterization via ultra-performance liquid chromatography-tandem mass spectrometry and in-vitro assays. *Journal of Chromatography*, 1217: 2512–2520.

Huang, H.W., C.P. Hsu, B.B. Yang, and Wang, C.Y. 2013. Advances in the extraction of natural ingredients by high pressure extraction technology. *Trends in Food Science and Technology*, 33: 54–62.

Ibáñez, E., A. Kubatova, F.J. Señoráns, S. Cavero, G. Reglero, and S.B. Hawthorne. 2003. Subcritical water extraction of antioxidant compounds from rosemary plants. *Journal of Agricultural and Food Chemistry*, 51: 375–382.

Ibáñez, E.A. and Oca, G., De-Murga, S., Lopez-Sebastian, J. Tabera, and G. Reglero. 1999. Supercrtical fluid extraction and fractionation of different preprocessed rosemary plants. *Journal of Agricultural and Food Chemistry*, 47: 1400–1404.

Irving, D. 2012. We are already up-scaling. *VMT Food Technology*, 16: 11–13.

Janositz, A. and D. Knorr. 2010. Microscopic visualization of pulsed electric field induced changes on plant cellular level. *Innovative Food Science and Emerging Technologies*, 11: 592–597.

Japon-Lujana, R. and M.D. Luque-de-Castro. 2006. Superheated liquid extraction of oleuropein and related biophenols from olive leaves. *Journal of Chromatography*, 1136: 185–191.

Jun, X. 2009. Caffeine extraction from green tea leaves assisted by high pressure processing. *Journal of Food Engineering*, 94(1): 105–109.

Kaufmann, B. and P. Christen. 2002. Recent extraction techniques for natural products: Microwave assisted extraction and pressurised solvent extraction. *Phytochemical Analysis*, 13: 105–113.

Khandare, V., S. Walia, M. Singh, and C. Kaur. 2010. Black carrot (Daucuscarota ssp. sativus) juice: Processing effects on antioxidant composition and color. *Food Bioproducts Processing*, 89(4): 482–486.

Koubaa, M., F. Barba, D. BursaćKovačević, P. Putnik, M.S. Santos, R.P. Queiros, S.A. Moreira, R.R. Duarte, and J.A. Saraiva. 2017. Pulsed electric field processing of different fruit juices. In: Rajauria, G., Tiwari, B. (eds.), *Fruit Juices*, 1st edn. Amsterdam: Elsevier, p. 760.

Kovačević, D.B., F.J. Barba, D. Granato, C.M. Galanakis, Z. Herceg, V. Dragović-Uzelac, and P. Putnik. 2018. Pressurized hot water extraction (PHWE) for the green recovery of bioactive compounds and steviol glycosides from Stevia rebaudiana Bertoni leaves. *Food Chemistry*, 254: 150–157.

Kovačević, D.B., P. Putnik, S. Pedisic, D. Jezek, S. Karlovic, and V. Dragović-Uzelac. 2015. High hydrostatic pressure extraction of flavonoids from freeze-dried red grape skin as winemaking by-product. *Annals ofNutrition and Metabolism*, 67: 521–522.

Lam, K. 2007. New aspects of natural products in drug discovery. *Trends in Microbiology*, 15(6): 279–289.

Leadbeater, N.E. 2014. Organic synthesis using microwave heating: Reference module in chemistry, molecular sciences and chemical engineering. *Comprehensive Organic Synthesis II*, 9: 234–286.

Li, B., B. Smith, and M. Hossain. 2006. Extraction of phenolics from citrus peels: I. Solvent extraction method. *Separation and Purification Technology*, 48(2): 182–188.

Li, H., Z. Deng, T. Wu, R. Liu, S. Loewen, and R. Tsao. 2011. Microwave-assisted extraction of phenolics with maximal antioxidant activities in tomatoes. *Food Chemistry*, 130(4): 928–936.

Li, Y., A.S. Fabiano-Tixier, M.A. Vian, and F. Chemat. 2013. Solvent-free microwave extraction of bioactive compounds provides a tool for green analytical chemistry. *TrAC Trends in Analytical Chemistry*, 47: 1–11.

Luque-de-Castro, M.D. and L.E. García-Ayuso. 1998. Soxhlet extraction of solid materials: An outdated technique with a promising innovative future. *Analytica chimica acta*, 369(1–2): 1–10.

Machado, A.P. D.F., J.L. Pasquel-Reátegui, G.F. Barbero, and J. Martínez. 2015. Pressurized liquid extraction of bioactive compounds from blackberry (*Rubusfruticosus L.*) residues: A comparison with conventional methods. *Food Research International*. 77: 675–683.

Mchugh, T. and S. Toepfl. 2016. Pulsed electric field processing for fruits and vegetables. *Food Technology*, 70: 73–75.

Miron, T.L., M. Plaza, G. Bahrim, E. Ibáñez, and M. Herrero. 2011. Chemical composition of bioactive pressurized extracts of Romanian aromatic plants. *Journal of Chromatography A*, 1218(30): 4918–4927.

Montesano, D., G. Rocchetti, P. Putnik, and L. Lucini. 2018. Bioactive profile of pumpkin: An overview on terpenoids and their health-promoting properties. *Current Opinion in Food Science*, 22: 81–87.

Moore, J., Z. Cheng, L. Su, and L. Yu. 2006. Effects of solid-state enzymatic treatments on the antioxidant properties of wheat bran. *Journal of Agricultural and Food Chemistry*, 54(24): 9032–9045.

Moreira, M.M., S. Morais, A.A. Barros, C. Delerue-Matos, and L.F. Guido. 2012. A novel application of microwave-assisted extraction of polyphenols from brewer's spent grain with HPLC-DAD-MS analysis. *Analytical and Bioanalytical Chemistry*, 403(4): 1019–1029.

Mortazavi, S.V., M.H. Eikani, H. Mirzaei, M. Jafari, and F. Golmohammad. 2010. Extraction of essential oils from Buniumpersicum Boiss using superheated water. *Food and Bioproducts Processing*, 88: 222–226.

Muller, G., W. Frey, M. Sack, C. Schultheiss, H.G. Mayer, J. Sigler, M. Kern, and U. Gunther. 2007. Karlsruhe electroporation system KEA: The success story of a technology transfer in the industry. *Research Center Karlsruhe-Nachrichten*, 39: 153–158.

Musina, O., P. Putnik, M. Koubaa, F.J. Barba, R. Greiner, D. Granato, and S. Roohinejad. 2017. Application of modern computer algebra systems in food formulations and development: A case study. *Trends in Food Science & Technology*, 64: 48–59.

Naczk, M. and F. Shahidi. 2004. Extraction and analysis of phenolics in food. *Journal of Chromatography A*, 1054(1–2): 95–111.

Oroian, M. and I. Escriche. 2015. Antioxidants: Characterization, natural sources, extraction and analysis. *Food Research International*, 74: 10–36.

Pasquet, V., J.R. Cherouvrier, F. Farhat, V. Thiery, J.M. Piot, J.B. Berard, R. Kaas, B. Serive, T. Patrice, and J.P. Cadoret. 2011. Study on the microalgal pigments extraction process: Performance of microwave assisted extraction: Process performance of microwave assisted extraction. *Process Biochemistry*, 46(1): 59–67.

Petigny, L., S. Périno, M. Minuti, F. Visinoni, J. Wajsman, and F. Chemat. 2014. Simultaneous microwave extraction and separation of volatile and non-volatile organic compounds of boldo leaves: From lab to industrial scale. *International Journal of Molecular Sciences*, 15(5): 7183–7198.

Plaza, M., M. Amigo-Benavent, M.D. del Castillo, E. Ibáñez, and M. Herrero. 2010a. Neoformation of antioxidants in glycation model systems treated under subcritical water extraction conditions. *Food Research International*, 43(4): 1123–1129.

Plaza, M., S. Santoyo, L. Jaime, G.G. B. Reina, M. Herrero, F.J. Señoráns, and E. Ibáñez. 2010b. Screening for bioactive compounds from algae. *Journal of Pharmaceutical and Biomedical Analysis*, 51(2): 450–455.

Pronyk, C. and G. Mazza. 2009. Design and scale-up of pressurized fluid extractors for food and bioproducts. *Journal of Food Engineering*, 95(2): 215–226.

Putnik, P., D. BursaćKovačević, K. Herceg, and B. Levaj. 2016. Influence of respiration on predictive microbial growth of Aerobic mesophilic bacteria and Enterobacteriaceae in fresh-cut apples packaged under modified atmosphere. *Journal of Food Safety*, 37(1): e12284.

Putnik, P., D. BursaćKovačević, A. RezekJambrak, F. Barba, G. Cravotto, A. Binello, J. Lorenzo, and A. Shpigelman. 2017a. Innovative "green" and novel strategies for the extraction of bioactive added value compounds from citrus wastes – A review. *Molecules*, 22(5): 680.

Putnik, P., D. BursaćKovačević, K. Herceg, S. Roohinejad, R. Greiner, A.E.D.A. Bekhit, and B. Levaj. 2017b. Modelling the shelf-life of minimally-processed fresh-cut apples packaged in a modified atmosphere using food quality parameters. *Food Control*, 81: 55–64.

Putnik, P., D. BursaćKovačević, K. Herceg, I. Pavkov, Z. Zorić, and B. Levaj. 2017c. Effects of modified atmosphere, anti-browning treatments and ultrasound on the polyphenolic stability, antioxidant capacity and microbial growth in fresh-cut apples. *Journal of Food Process Engineering*, 40(5): e12539.

Putnik, P., D. BursaćKovačević, D. Jezek, I. Sustic, Z. Zorić, and V. Dragović-Uzelac. 2017d. High-pressure recovery of anthocyanins from grape skin pomace (Vitis vinifera cv. Teran) at moderate temperature. *Journal of Food Processing and Preservation*, 42(1): e13342.

Putnik, P., S. Roohinejad, R. Greiner, D. Granato, A.E.D.A. Bekhit, and D. Bursac Kovacevic. 2017e. Prediction and modeling of microbial growth in minimally processed fresh-cut apples packaged in a modified atmosphere: A review. *Food Control*, 80: 411–419.

Putnik, P., F.J. Barba, J.M. Lorenzo, D. Gabric, A. Shpigelman, G. Cravotto, and D. BursaćKovačević. 2017f. An integrated approach to mandarin processing: Food safety and nutritional quality, consumer preference and nutrient bioaccessibility. *Comprehensive Reviews in Food Science and Food Safety*, 16: 1345–1358.

Putnik, P., F.J. Barba, I. Spanic, Z. Zorić, V. Dragović-Uzelac, and D. BursaćKovačević. 2017g. Green extraction approach for the recovery of polyphenols from Croatian olive leaves (Olea europea). *Food and Bioproducts Processing*, 106: 19–28.

Rajha, H.N., N. Boussetta, N. Louka, R.G. Maroun, and E. Vorobiev. 2015. Effect of alternative physical pretreatments (pulsed electric field, high voltage electrical discharges and ultrasound) on the dead-end ultra-filtration of vine-shoot extracts. *Separation and Purification Technology*, 146: 243–251.

Robbins, R.J. 2003. Phenolic acids in foods: An overview of analytical methodology. *Journal of Agricultural and Food Chemistry*, 51(10): 2866–2887.

Rosello-Soto, E., M. Koubaa, A. Moubarik, R.P. Lopes, J.A. Saraiva, N. Boussetta, and F.J. Barba. 2015. Emerging opportunities for the effective valorization of wastes and by-products generated during olive oil production process: Non-conventional methods for the recovery of high-added value compounds. *Trends in Food Science and Technology*, 45(2): 296–310.

Santos, D.T., P.C. Veggi, and M.A. A. Meireles. 2012. Optimization and economic evaluation of pressurized liquid extraction of phenolic compounds from jabuticaba skins. *Journal of Food Engineering*, 108(3): 444–452.

Shi, J., C. Yi, S.J. Xue, Y. Jiang, Y. Ma, and D. Li. 2009. Effects of modifiers on the profile of lycopene extracted from tomato skins by supercritical CO_2. *Journal of Food Engineering*, 93(4): 431–436.

Simsek, M., G. Sumnu, and S. Sahin. 2012. Microwave-assisted extraction of phenolic compounds from sour cherry pomace. *Separation Science and Technology*, 47(8): 1248–1254.

Starmans, D.A. and H.H. Nijhuis. 1996. Extraction of secondary metabolites from plant material: A review. *Trends in Food Science & Technology*, 7(6): 191–197.

Szentmihályi, K., P. Vinkler, B. Lakatos, V. Illés, and M. Then. 2002. Rose hip (Rosa canina L.) oil obtained from waste hip seeds by different extraction methods. *Bioresource Technology*, 82(2): 195–201.

Tao, Y., D.W. Sun, A. Górecki, W. Błaszczak, G. Lamparski, R. Amarowicz, and T. Jeliński. 2016. A preliminary study about the influence of high hydrostatic pressure processing in parallel with oak chip maceration on the physicochemical and sensory properties of a young red wine. *Food Chemistry*, 194: 545–554.

Toepfl, S., A. Mathys, V. Heinz, and D. Knorr. 2006. Review: Potential of high hydrostatic pressure and pulsed electric fields for energy efficiency and environmentally friendly food processing. *Food Review International*, 22: 405–423.

Vergara-Salinas, J.R., P. Bulnes, M.C. Zúñiga, J. Pérez-Jiménez, J.L. Torres, M.L. Mateos-Martín, and J.R. Pérez-Correa. 2013. Effect of pressurized hot water extraction on antioxidants from grape pomace before and after enological fermentation. *Journal of Agricultural and Food Chemistry*, 61(28): 6929–6936.

Verma, S.K., P. Goswami, R.S. Verma, R.C. Padalia, A. Chauhan, V.R. Singh, and M.P. Darokar. 2016. Chemical composition and antimicrobial activity of bergamot-mint (*Mentha citrataEhrh.*) essential oils isolated from the herbage and aqueous distillate using different methods. *Industrial Crops and Products*, 91: 152–160.

Villa-Rodríguez, J.A., F.J. Molina-Corral, J.F. Ayala-Zavala, G.I. Olivas, and G.A. González-Aguilar. 2011. Effect of maturity stage on the content of fatty acids and antioxidant activity of 'Hass' avocado. *Food Research International*, 44(5): 1231–1237.

Vinatoru, M. 2001. An overview of the ultrasonically assisted extraction of bioactive principles from herbs. *Ultrasonics Sonochemistry*, 8(3): 303–313.

Wang, L. and C.L. Weller. 2006. Recent advances in extraction of nutraceuticals from plants. *Trends in Food Science & Technology*, 17(6): 300–312.

Wang, L., C.L. Weller, V.L. Schlegel, T.P. Carr, and S.L. Cuppett. 2008. Supercritical CO_2 extraction of lipids from grain sorghum dried distillers grains with solubles. *Bioresource Technology*, 99(5): 1373–1382.

Wang, T., R. Jonsdottir, H.G. Kristinsson, G.O. Hreggvidsson, J.O. Jonsson, G. Thorkelsson, and G. Olafsdottir. 2010. Enzyme-enhanced extraction of antioxidant ingredients from red algae Palmariapalmata. *LWT - Food Science and Technology*, 43(9): 1387–1393.

Wu, J.J., J.C. Lin, C.H. Wang, T.T. Jong, H.L. Yang, S.L. Hsu, and J.C. Chieh-ming. 2009. Extraction of antioxidative compounds from wine lees using supercritical fluids and associated anti-tyrosinase activity. *The Journal of Supercritical Fluids*, 50(1): 33–41.

Xu, F.X., C. Yuan, J.B. Wan, R. Yan, H. Hu, S.P. Li, and Q.W. Zhang. 2015. A novel strategy for rapid quantification of, 20(S)-protopanaxatriol and, 20(S)-protopanaxadiol saponins in Panax notoginseng P. ginseng and P. quinquefolium. *Natural Product Research*, 29(1): 46–52.

Xu, Y., F. Cai, Z. Yu, L. Zhang, X. Li, Y. Yang, and G. Liu. 2016. Optimisation of pressurised water extraction of polysaccharides from blackcurrant and its antioxidant activity. *Food Chemistry*, 194: 650–658.

Yahya, A. and R.M. Yunus. 2013. Influence of sample preparation and extraction time on chemical composition of steam distillation derived patchouli oil. *Procedia Engineering*, 53: 1–6.

Yi, C., J. Shi, S.J. Xue, Y.M. Jiang, and D. Li. 2009. Effects of supercritical fluid extraction parameters on lycopene yield and antioxidant activity. *Food Chemistry*, 113: 1088–1094.

Yi, Y., Q.W. Zhang, S.L. Li, Y. Wang, W.C. Ye, J. Zhao, and Y.T. Wang. 2012. Simultaneous quantification of major flavonoids in "Bawanghua", the edible flower of Hylocereusundatus using pressurised liquid extraction and high performance liquid chromatography. *Food Chemistry*, 135(2): 528–533.

Zhang, H.F., X.H. Yang, and Y. Wang. 2011. Microwave assisted extraction of secondary metabolites from plants: Current status and future directions. *Trends in Food Science and Technology*, 22: 672–688.

Zhang, H.F., X.H. Yang, L.D. Zhao, and Y. Wang. 2009. Ultrasonic-assisted extraction of epimedin C from fresh leaves of Epimedium and extraction mechanism. *Innovative Food Science and Emerging Technologies*, 10: 54–60.

Zhang, Z. and Z. Su. 2002. Catalysis mechanism to increase taxol from the extract of Taxus cuspidate callus cultures with alumina chromatography. *Separation Science and Technology*, 37(3): 733–743.

Zill-e-Huma, M.A. Vian, A.S. Fabiano-Tixier, M. Elmaataoui, O. Dangles, and F. Chemat. 2011. A remarkable influence of microwave extraction: Enhancement of antioxidant activity of extracted onion varieties. *Food Chemistry*, 127(4): 1472–1480.

Index

A

B

C

D

E

F

H

I

K

U

V

W

Z